THE
ABILITY
HACKS

By Greg Shaw

First published in 2018 by Microsoft Corporation
One Microsoft Way
Redmond, Washington 98052

©2018 Microsoft. All rights reserved.

ISBN 9781983089978

FOREWORD

By Peter Lee

A few years ago, Microsoft engineer Jay Beavers (who you will read more about in this book), along with several others, rolled into my office in a motorized wheelchair controlled by the gaze of his eyes. I should have been thrilled. After all, I was just starting on a difficult assignment to create a new breed of product-engineering team that might be more proficient at harnessing the enormous firepower of Microsoft Research. I was desperate to find researchers with promising ideas, the thought being that I could attempt to get from concept to product as quickly as possible by investing engineering, design and program-management resources around them. Jay was exactly the type of person I was looking for. And he just dropped – or rolled – right into my office.

Now, as you read this book it will sound a bit like I heroically jumped at the chance to invest in the wheelchair project. But, in fact, my initial reaction, which I tried hard to hide from the team, was one of annoyance. OK, I thought, here we have a hackathon-winning project with a feel-good story involving a famous athlete, Steve Gleason. Great. But is this a team motivated more by PR than by real-world impact? I was frankly suspicious of their motives. And how on earth would we get something like this to market? Microsoft would almost certainly never get into the business of selling wheelchairs (even AI-powered ones). And finally, I was getting tired of seeing pitches involving lots of demos but lacking any grounding in a solid business concept. I had lost interest in investing in fancy demos. I wanted new products with a real chance for sustainable real-world impact.

In my grumpiness, I kept a forced smile and was on the verge of politely telling Jay, and the others who came with him, the deadly "I will think about it and get back to you, thank you." But then, somehow, the discussion turned. Someone brought up unforeseen shortcomings in the system architecture that became apparent once the team started working with Steve Gleason. Another person started to explain concepts in user-centered design, inclusive design, universal design, and what these implied for software design and engineering. Jay described the rapid feedback cycle of design <->engineering<->real-world use and what this meant for Windows and our development stack. And as the discussion wore on, it became more obvious that maybe there was actually something special going on here, something that was authentically devoted to empowering people.

The whole encounter took no more than 10 or 15 minutes, but ultimately was as convincing as any 40-minute whizzy PowerPoint-powered project pitch could ever be. It had several crucial moments and, in the end, led to the decision to fund an initial productization effort, with the first requirement being to define objectively assessable "impact goals" and the concrete plan for getting there. My colleague, Rico Malvar, kindly took on the task of guiding this process, growing the team, connecting it across the company and, ultimately, instilling solid product-engineering discipline while at the same time accommodating the free-flowing mindset of research. Other people like Ann Paradiso, who had critical design skills that we needed, were cajoled over beers at a local pub and reassured that joining this unusual team wouldn't be a career-stifling move.

Now, several years later, key technologies from this project, such as Eye Control, are standard features in Windows 10, and the path to new, more natural and accessible ways of interacting with computers, such as speech, mobility and other skills, is clearer than ever. The process has brought us closer to technology partners inside and outside of Microsoft. Even more importantly, it has gotten us tightly integrated with teams of PALS – People with ALS – who have worked so hard with us to make the technology real. As you might imagine, all of this gives us feelings of tremendous satisfaction. But we also understand keenly that we – and the Microsoft development community at large – still have a huge amount of work to do. Looking back on this episode, it was one of the most important growth experiences for me personally and contributed fundamentally to the culture of the New Experiences in

Technology (NExT) organization that has grown up within Microsoft since then and helped to influence a number of projects born from Microsoft Research.

The other major story in this book, on Learning Tools, involves a different set of players in a different part of Microsoft, and was born in a different set of circumstances. But in what I think of as the most relevant characteristics, the story of Learning Tools shares the same lessons and opportunities for personal and organizational growth. I know through my interactions with several of that team's members that we have all begun to internalize into our engineering cultures the seemingly simple but ultimately subtle idea: a focus on inclusion helps a team become more empathetic with its users, which in turn affects deeply the design and development process of products. Inclusion injects crucial energy into rapid feedback cycles that is core to innovation. Inclusive design also creates better products. In the same way the Eye Control project has affected the innovation process in Microsoft Research and NExT, I can see that Learning Tools is having an effect on the culture in the Microsoft Office and Windows teams.

The final part of this book is important to all of us at Microsoft, though we recognize it only as the start of a journey to innovate through empowerment. This part presents a catalog of some of our existing accessibility features and technologies. The catalog is humble and not exhaustive, but at least for us, putting it together was an important exercise because it forces a change in perspective. By looking at our entire software development stack through the lens of inclusion, and sharing with all of you, we start down the path of conceptualizing software development differently.

My colleague, Jenny Lay-Flurrie, uses the phrase "innovating through ability and innovating through disability." This seems so apt now, because reflecting on the contents of this book, I have come to learn an important lesson about the nature of innovation. Often, when we think about technological innovation, we think about technological disruption. But for me, at least, the main lesson is that innovation comes more from an intent to empower than from an intent to disrupt. We hope you will be motivated to contribute and embark with us on this journey of empowerment and inclusion together.

There was a palpable sense of this being important. Microsoft should be doing this. We have the people and the resources. That was the main vibe.

-Microsoft engineer on building an eye-controlled wheelchair.

PREFACE

This is the story of two Microsoft hackathon teams, one in the summer of 2014 and one the following summer of 2015. The first would pioneer new software to revolutionize the mobility of tens of thousands of people who live with severe paralysis caused by ALS, Parkinson's, cerebral palsy and traumatic neurological injuries. The second team would pioneer software to help kids with dyslexia read and love learning for the first time in their lives. It's the story of two small groups of driven, focused and passionate software engineers, program managers, marketers and advocates. It's the story of realizing the transformative power of technology for people with disabilities, not just for traditional consumer and industrial markets. It's the story of ignoring the company's and the industry's checkered past in making inclusive technology, and doing something truly great — improving outcomes for everyone, discovering a design ethos and blazing a new trail for accessibility.

More than one billion people around the world live with a disability of some kind, and it's estimated two-thirds of us know someone with a disability. This book explores an optimistic belief that computer software and hardware can empower people with disabilities in a multitude of scenarios.

The language used within and about this important community is not without controversy. Disability is not a term we shy away from, though it is imperfect. In their 2005 writing collection, *Beyond Victims and Villains: Contemporary Plays by Disabled Playwrights,* the contributors were united in rejecting euphemistic terms for disability and "universalizing stories or narratives of disability, and [united] in their refusal to accept the subtext of superiority that underlies the charitable gesture." None of the writers in this anthology identify as "physically challenged," "differently abled" or "handi-capable," the terms for disability so easily satirized in popular culture. In the early 1970s, a group of young people with significant disabilities, calling themselves the "Rolling Quads," decided to throw off the invisibility cloak of shame and reclaim the negative term "disability" as a banner of pride and power. The first generation of disability civil rights activists hoped for a similar rehabilitation of literary and social identity by self-consciously reclaiming the terms "disability" and "disabled." There were many important moments over the 20 years that led to the landmark federal action. Framers of the 1990 Americans with Disabilities Act (ADA) said a person is considered to have a disability if he or she:

- Has a physical or mental impairment that substantially limits one of more of the major life activities of such an individual

- Has a record of such an impairment
- Is regarded as having such an impairment

Many today, however, reject the term impairment. The World Health Organization (WHO), for example, made the following statement:

> Disabilities is an umbrella term, covering impairments, activity limitations, and participation restrictions. An impairment is a problem in body function or structure; an activity limitation is a difficulty encountered by an individual in executing a task or action; while a participation restriction is a problem experienced by an individual in involvement in life situations.

> Disability is thus not just a health problem. It is a complex phenomenon, reflecting the interaction between features of a person's body and features of the society in which he or she lives. Overcoming the difficulties faced by people with disabilities requires interventions to remove environmental and social barriers.

> Almost everyone will be temporarily or permanently impaired at some point in life, and those who survive to old age will experience increasing difficulties in functioning, according to the WHO.

Cliff Kuang at Co.Design once wrote that we "stand at the end of a long line of inventions, which might have never existed, but for the disabled." In 1808, Pellegrino Turri built the first typewriter so his blind lover could write letters more legibly. In 1876 Alexander Graham Bell invented the telephone to support his work helping the deaf. And in 1972, Vint Cerf programmed the first email protocols for the nascent Internet. Email was the only seamless way to communicate with his wife, who was deaf, while he was at work.

As one engineer interviewed for this project said, "It's not about the technology. It's about the people."

Part I

TEAM GLEASON

Prelude to a hack

Around the world, 2014 began with the glow of unity that rolls around every four years with the Winter Olympics, that year held in Sochi, Russia. But within weeks a complicated world became more complex as Russia annexed Crimea, Ebola spread, ISIS arose as a new terrorist threat, and tensions heightened between police and the African American community in cities across the U.S. In the tech sector, evidence pointed to North Korea as perpetrator of a malicious hack of Sony Pictures. And a little-known leader within Microsoft, Satya Nadella, was named as the company's third CEO.

In February, a newly refreshed Microsoft aired its

first ever Super Bowl ad when hometown favorites, the Seattle Seahawks, took on the Denver Broncos. The ad was widely heralded as emotionally moving and focused on empowerment rather than the more customary portrayal of consumer and enterprise productivity features. The ad highlighted Steve Gleason, the former New Orleans Saints safety and advocate for people with disabilities. In 2006, on the night the Superdome reopened following the horrors of Hurricane Katrina, Steve blocked an Atlanta Falcons punt that was converted into a touchdown. A statue of his football heroics outside the Superdome would eventually commemorate the city's "Rebirth." Steve, a native of Spokane, Washington, became a New Orleans hero. In 2011, Steve revealed he was living with ALS, a neurological disease that left him with no mobility other than the movement of his eyes and robbed him of his voice. It did not, however, take away his will to ignite change, and only strengthened his belief that technology could make his physical life – and that of many others -- better.

During the Seahawks-Broncos Super Bowl, the millions who tuned in saw Steve use his eyes and a software program to type the ad's powerful script: What is technology? What can it do? How far can it take us? Technology inspires us, takes us places we've only dreamed of..."

It was an instant hit, ranking as the Super Bowl's number one ad by multiple news critics and the Kellogg

School of Business. "Microsoft's Super Bowl ad reminds the world why its software matters," read one headline.

Determined not to let the moment slip away into thin air, Steve resolved that night to use his new eye gaze technology to write yet another script that would rattle some cages back at the software mother ship in Redmond, Washington. Even with the new technology, which tracks his eye movements to a keyboard and translates his intentions into typing, Steve can write just 15 to 20 words per minute. By comparison, the late professor Stephen Hawking, the Cambridge theoretical astrophysicist, typed just several words per minute, according to news reports.

Steve gathered his thoughts, let some time go by, and then sent an email to Microsoft's head of corporate citizenship thanking the company for the visibility generated by the ad and the $25,000 Microsoft contributed to the Team Gleason foundation.

"But now we need to fix your technology," he wrote.

His email included a long list of bugs that affected people with disabilities. People who wear glasses find it difficult to use eye gaze. "Swiping" was darn near impossible using your eyes. The on/off button was difficult to use. He wanted to be able to talk more naturally with his wife. He wanted the ability to play with his young son. At the very end of his email, which took hours to compose, he said he wanted to be able to drive his wheelchair with the one thing he could still move, his eyes.

Unfortunately, the email languished somewhere in the vast network of corporate and business leaders who could do something about his feedback. It was a time of organizational and cultural change at the company. Steve never received a reply.

Gearing Up

Late in May, Steve traveled to New York City for the Social Innovation Summit, described on its website as a convening that plays "at the nexus of technology, investment, philanthropy, international development and business to investigate solutions and catalyze inspired partnerships that are disrupting history." There in the halls and conference rooms of the United Nations Plaza, tech companies, bankers and foundation leaders listened to speeches and schemed about what could be. Microsoft's then-head of global citizenship, Akhtar Badshah, was in the audience when Steve gave an inspiring speech about how his life is dependent on technology and how he wanted to be able to do more. After the speech, Akhtar met Steve and learned about his unanswered email. He promised to investigate.

As luck would have it, Akhtar was having dinner with Satya that Friday night back in Seattle. At some point during the meal, the new CEO turned to Akhtar and asked what he was up to. Akhtar talked about the company's giving, and then mentioned something that piqued Satya's interest: Steve, the star of the Super Bowl ad, had ideas for how to continue to improve the technology. As part of Microsoft' cultural refresh, Satya had decided to jettison the traditional annual company meeting and instead sponsor an all-Microsoft hackathon. What if Steve came to campus later that summer to help with a hack that could benefit himself and thousands of others like him?

Akhtar wasted no time. He later sat down at his desk in Building 8 to pound out an email on his Surface Pro to Satya and his chief of staff. Akhtar relayed that he was now in touch with Steve and, at Satya's offering, would fly him and his support team to the Redmond campus. Akhtar was excited, but behind his email was the overwhelming sense of not quite knowing where to start.

He didn't have to wait long. An hour and a half later, Satya fired back an email adding Jenny Lay-Flurrie, who chaired the disability employee community, and Dave Campbell, a top engineer from the Cloud and Enterprise team. Jenny reached out to Steve and the nonprofit he had founded, Team Gleason, as leaders of a similar project that had kicked off a year earlier for the blind (later released as the project Microsoft Soundscape).

"I don't think this would have gone anywhere if Satya hadn't gotten involved," Akhtar later recalled.

Satya wrote in his 2017 book, *Hit Refresh*, about he and his wife Anu's son, Zain, who has severe cerebral palsy and has been in a wheelchair all his life. Making technology accessible to everyone would become a hallmark of his executive leadership.

When Satya's email landed in Jenny's inbox, she recognized its significance. This story already had national attention, the new boss was asking for her help and she was insanely curious about eye gaze technology. She already had nine projects lined up for the Ability Hack and quickly placed a call to Team Gleason to introduce herself and to follow up. She also posted a notice internally at Microsoft asking for help and suggesting a team be spun up in advance of the upcoming hackathon. It was June and the hackathon was scheduled for end July.

Jenny would eventually ask Matthew Mack from her team, who had never led a hackathon before, to lead the day-to-day operations of the Gleason hack. That week, they sent out a call to arms, an invitation to join one of 10 hacks including Steve's eye gaze challenge. "Using feedback from the disability community and Steve Gleason, this hack is focused on enabling a Surface 3 to be powered on/off with eye gaze. While that sounds simple, this is a nut that has not been cracked by the industry and would set us above the competition. It's also at the TOP of Steve Gleason's list." She asked for help from those who knew Surface hardware, cameras and visual recognition, the Windows operating service and device engineering.

One objective was to enable third party tech companies like Tobii, a Swedish firm that leads in the development of eye control and eye tracking, to bring to market eye gaze functionality for the disabled community. Not only would this help Steve and others with limited mobility, but it would become a competitive advantage for Microsoft by improving Windows and Windows devices.

Since the 1990s, Microsoft has focused increasingly on accessibility. In the 20-plus years since, there have been moments of brilliance and moments of learning. During that time, Microsoft had fallen in and out of favor. In the early days, the blind community liked Microsoft's DOS operating system, but the Windows graphical user interface (GUI) suddenly made computers much less accessible. "Blind Users Feel Abandoned as Computers Shift to Icons," one headline complained in the mid-1990s. Over the years, Microsoft introduced a range of remedies, including StickyKeys, Active Accessibility APIs for developers and more. Internally, disability employee groups were growing, some dating back to the '90s. In 2011, there were six groups for people with blindness, deafness, mobility and other disabilities. The company hosted its first employee "Ability Summit" in 2010. Eighty people attended.

"Accessibility was not part of the priority," said

Jenny, a talented product manager who hailed from near Birmingham, England and who had for years covered for her own deafness before seeing it as a strength. That strength and her talents ultimately gained her a reputation within the company for effective accessibility advocacy and passionate leadership.

"I joined every disability alias I could when I started out at Microsoft," Jenny recalled. "I stalked and listened. They were all saying the same things: I don't know how to tell my manager about my disability. Is Microsoft a company for me? Aren't we a technology company, can't we do something?"

She stopped lurking and started joining each disability group one-by-one, unifying them and eventually chairing the unified group Disability All.

"We lost our way at times."

As Seattle's rainy "June-uary" turned into a spectacular Pacific Northwest July, word was getting around campus about the hackathon. Jennifer Zhang, a program manager, had never done a hackathon before and decided to focus on the ability hacks.

"I chose the eye gaze project because it was the most compelling," she said. "It was the most personally compelling for me. I have a friend with muscular dystrophy. I've known him since junior year of high school and I've seen the drastic steady decline since then. He's thankfully outliving all expectations he was initially told he could have of lifespan. Seeing his struggles with technology really were what drew me to become a subject matter expert in accessibility and what drew me to being part of this project."

Daniel Deschamps, an engineer focusing on rapid hardware development, was at home on paternity leave with a newborn baby when he saw an email with the list of hacks. He felt a surge of excitement when he saw the Gleason opportunity. He thought about the need to create ways to filter light so that it didn't interfere with eye control.

Shane Williams, an Australian engineer in Microsoft Research, lights up when he recalls the day Jennifer, Vidya Srinivasan from the OneDrive group and Matthew Mack, the energetic leader of the hack, descended on his office to twist his arm to join the project. Shane was an experienced eye control software developer, and the emerging team was still thin on deep technical know-how.

"Matthew's energy level is quite high," Shane explained. "He's an excitable character and was persuasive. They came to my office, and it was hard not to get involved. We had been working on eye tracking for some time and we wanted it to go places, so this seemed like a good business proposition, as well."

Eyes on the Prize

Within weeks, the core of the Team Gleason hack group was in place. The hackathon would commence on July 29, so time was of the essence. A series of "think tank" sessions began with the goal of imagining what was possible and how to get it done.

The initial focus remained on Steve's complaint about the Surface shutting down on its own in the midst of his daily activities, which rendered the entire eye control and tracking system useless. The power button required the user to touch the top of the monitor.

"Our primary investigation remains 'eye gaze on,' which is evolving to 'Surface always on.' Rather than looking at the problem of how to run on the Surface the team has taken the approach of never letting the device shut down," Matthew wrote in one think tank summary.

A range of options were being explored from updating the UEFI firmware to updating Windows in a way that would make the restart process simpler. As always, battery life was an irritating design constraint.

Meanwhile, over in the Xbox building, Jon Campbell was working with some interns from Washington State University. A software engineer born and raised in the Seattle area, Jon applied to Microsoft 12 times – once every semester for six years while studying computer science and math in college and grad school. He eventually started as a web designer for Microsoft, but on this particular day, he was designing and testing functionality for Kinect, the motion-sensing technology for videogames. Kinect was a commercial breakthrough for natural user interfaces, and later went on to become a key ingredient in robotics. Jon was talking with the interns when their WSU professor mentioned that Steve Gleason, who had played football in college for the Cougars, was working with some people at Microsoft. Perhaps Jon would like to get involved.

"As a developer I never met customers," Jon said, sitting inside his crowded research lab. "Your job is to be at the desk coding. I really wanted that experience, to say, there is the customer right there."

The eye gaze hack team had already gotten started on some of the items in Team Gleason's email, but Jon reached out and got himself invited to one of their think tank sessions. And it was an eye opener.

"It was a terrible meeting – a real train wreck," Jon said, shaking his head still in disbelief.

After an hour of discussion and very little progress, they pulled up Steve's email again, which Jon had not seen before. He read over the list of items to be fixed when he suddenly sat up straight in his chair at the last idea in the email.

"I saw 'drive wheelchair with eyes.' I thought, wait a

minute. Kinect is not good at detecting eye movements, but what if we turned it the other way? What if Kinect sat on his head and looked at a keyboard?"

Matthew sensed opportunity and pulled Jon into an after-meeting to brainstorm. And therein lies the spirit of hacking. Matthew and Jon discovered that Kinect had the necessary hardware to be an eye tracker; specifically it has an infrared-sensitive (IR) webcam and an IR emitter. Back then, the Kinect version 1.0 was starting to show up at Gamestops and other outlets for $20 each, so it was a super attractive option to see if it could be used as an eye tracker. At that time, eye trackers could cost tens of thousands of dollars.

Jon had a Kinect version 2.0 and felt the latest iteration might be made to work as an eye tracker. But that idea fell flat since, ultimately, they would use a commercial tracker like the Tobii EyeX.

"In that meeting, I sat there and thought, well, OK, we don't want to repurpose the Kinect as an eye tracker, but suppose we had an eye tracker driving a wheelchair robotically? Lots of people use the Kinect for robotic navigation and object avoidance. What if instead of trying to have the Kinect face the person and then write eye tracking software, which is a lot of work that we don't have time for, what if instead we had the Kinect face forward and basically thought of the chair like a robotics problem where there is a human pilot. If that is the case, then we could use the Kinect like the robotics folks do – for object avoidance and navigation."

The ingredients for the hack were beginning to fall into place: Windows and the Surface Pro, which had been launched the prior year, Kinect, predictive texting, an Arduino UNO microcontroller, the Tobii eye control software developer kit and a 300-pound Permobil c500 wheelchair.

But rather than jump right into building, the team took an important first step. They wouldn't just fall into old habits; they needed, in essence, to put themselves in the dark. The Gleason hackathon team agreed to begin with what they called "design empathy." They searched to find insights that could only come from putting themselves in Steve's place, sitting in a wheelchair and imagining. They counted down the hours till Steve arrived on site; they needed his expert insight.

Around the same time, Kat Holmes had joined Microsoft as principal director of what was being called Inclusive Design. She pioneered the development of the Microsoft Inclusive Design toolkit, which Fast Company would later describe as a radical evolution of design thinking and practices. Unlike traditional approaches to accessibility, Kat emphasized studying the way people interact with each other as an analog for better interactions between people and technology. Her aim was to create experiences that are *one size fits one*, not one size fits all.

"Inclusion can be a source of innovation and growth, especially for digital technologies," she writes in her book *Mismatch: How inclusion shapes design*. "It can be a catalyst for creativity and an economic imperative. There are many challenges that stand in the way of inclusion, the sneakiest of which are sympathy and pity. Treating inclusion as a benevolent mission increases the separation between people. Believing that it should prevail simply because it's the right thing to do is the fastest way to undermine its progress. To its own detriment, inclusion is often categorized as a feel-good activity."

According to inclusive design principles, software should be built to:

- Recognize exclusion. Exclusion happens when we solve problems using our own biases.
- Learn from human diversity. Human beings are the real experts in adapting to diversity.
- Solve for one, extend to many. Focus on what's universally important to all humans.

As team leader, Matthew met with Kat early on to inform the team's work. Vidya Srinivasan from the OneDrive group, a key program manager in the hack, brought experience with another design ethos – universal design. She had graduated from North Carolina State University's Human-Computer Interaction Lab. NC State pioneered universal design, which has seven principles: equitable use, flexibility in use, simple and intuitive use, perceptible information for the user, tolerance for error, low physical effort and, finally, appropriate size and space. It would be a stretch to say these design goals guided Team Gleason — after all, a hack is about getting things done quickly with what you have — but over time the wheelchair would embody inclusive and universal design aims.

In pursuit of equitable use, the team resolved to use big buttons so that even those who have less eye tracking precision could still use the app. By integrating functionality into the wheelchair manufacturer's control, they allowed for the flexibility of speed and agility that the wheelchair already provides. They wanted a simple design, something that could be taught in under a minute. It would be intuitive and straight forward to use. For perceptible information, they used a pass-through camera, like those utilized in virtual reality, so operators would always know where they were going even though their eyes might be concentrating on a keyboard. In terms of tolerance for error, their smoothing algorithms and user interface setup would allow the person in the wheelchair to drive even when traversing bumps. If it is so bumpy that eye tracking was lost, the wheelchair would stop for safety. The wheelchair would require minimal physical effort as it wouldn't require a button to be held to make the chair go, and it wouldn't require the user to stare at a single spot while driving. With pass-through video, the driver could keep his or her attention in a single space rather than having to stop the chair and look around constantly. The team optimized the app for this task, providing large hit targets and wide margins. The wheelchair driving app is integrated with the Augmentative and Alternative Communication (AAC) solution used to supplement or replace speech, thereby requiring minimal effort for switching tasks. The hardware for controlling the chair is small and can easily be tucked away on the chair such that it is not obtrusive.

The notion of universal design is controversial if you consider the notion that there is no "one size fits all."

Jutta Treviranus, Ontario College of Art and Design professor and director and founder of the Inclusive Design Institute, explains the distinction this way: "Universal design is one-size-fits-all. Inclusive design is one-size-fits-one. Inclusive design might not lead to universal designs. Universal designs might not involve the participation of excluded communities. Accessible solutions aren't always designed to consider human diversity or emotional qualities like beauty or dignity. They simply need to provide access. Inclusive design, accessibility and universal design are important for different reasons and have different strengths. Designers should be familiar with all three."

The team went to work hacking together the various components with the goal of getting a wheelchair to move at a computer's request, and of course for the computer to be directed by someone who cannot move his or her limbs, only the eyes. In the past, there were only two solutions for those with quadriplegia – a sip-and-puff tube or head switcher rays. Neither of those solutions worked for Steve Gleason, or for many people with severe limitations that come with disabilities like ALS.

In the days that followed, there were late nights, passionate discussions, compromises, successes and failures. The team members' goals were ambitious, and their time together was brief.

First, they imagined the Surface device "always on." Steve was dependent on another person to power on his machine. They would have to modify the UEFI code in the Surface 3 firmware to detect when the machine shuts down and automatically reboots without requiring any manual interaction. They actually had to change the Surface hardware to accommodate the hack's goals, according to Daniel Deschamps, the engineer.

Second, they wanted natural speech, the ability for Steve to use his eyes to type messages. This meant developing native AAC communication software removing the need to download third party software or improve third party applications. They wanted to enable users dependent on eye tracking keyboards to speak faster, more naturally, and synchronously using predictive text on Windows. They designed a custom keyboard suitable for eye tracking interaction and built a predictive text algorithm for Windows.

Third, and the toughest nut to crack, Steve wanted his independent mobility back. This meant enabling those who use an electric wheelchair and who do not have adequate motor control to navigate the wheelchair using eye tracking technology. As a first step, they built eye tracking interactive drive that uses Kinect V1 for obstacle detection and avoidance. To enhance this further, they built eye tracking autonomous drive by using the Robotics Autonomous Navigation and Suggestion Drive to navigate the wheelchair from one location to another by itself, while avoiding obstacles.

Deadline

On Wednesday, July 23, with three days before the kickoff of the hackathon competition, Matthew Mack sent one of his daily reports to the team just a little before 1 a.m. "Another amazing day." The team was making progress on the eye gaze technology after a conversation with the Windows Mobile operating system team to improve various work flow issues. He was pleased to announce that Shane Williams had decided to join the hack team after their persuasive visit to his office. Shane had helped them make progress with the product team. That night, 20 hours later, Matthew sent another update calling for more eye gaze brainstorming to work on Swype, the predictive keyboarding that would make it easier for Steve to type. He also began to set the stage for timing and plans for the actual showmanship that would be necessary to win the hackathon the following week – people's roles, a video and the itinerary.

The following day, Matthew was back at it. "Things are firming up." He quickly summarized the day's technical progress, and then turned to something he felt would be important to their success: the story they would tell. On the day of the hackathon his "story team" would videotape a demonstration and interviews with team members. The aim was to make sure anyone who saw the video online or at their booth would vote for Team Gleason in the hackathon.

On Friday, with just the weekend to go, Matthew reported on a breakthrough from the robotics team members. "Out of the box thinking resulted in a strategy to keep the Surface from never shutting down. This elegant solution addresses the ability to keep the device updated whilst ensuring that help is not needed to restart an (unwanted) shutdown that was central to eye gaze."

Despite progress, Matthew and the hack team worried they were not taking full advantage of Microsoft's deep expertise. Matthew had been introduced to Gershon Parent, a software engineer and roboticist, by a researcher in China, but Gershon was adamant he did not want to join another hack team. Matthew was known for not taking no for an answer and got him to agree to a meeting in Building 20 on the Saturday before the hackathon was scheduled to commence. They met and Gershon showed him a robot he was working on. Eventually Gershon softened as they discussed the possibilities for robotics and people with disabilities.

"Gershon was extremely outgoing and enthusiastic, bringing great energy to the team," Matthew recalled years later.

They were working on the mapping software that allows users to select a spot on a previously mapped area and send the wheelchair to that spot – an autonomous vehicle of sorts. Their plan was to simultaneously approach the wheelchair hack from two perspectives: to enable Steve to drive his wheelchair using his eyes and to automate a Pioneer Robot that would look like Steve's as a "scale model" just in case to show what's possible.

A big breakthrough came when Shane was able to coalesce the wheelchair driver's eye gaze into X and Y coordinates that could then push buttons. This required careful calibration. Better trackers can follow eye movements within a half degree, or about the width of a thumb when viewed at an arm's length.

On Monday of the hackathon, the wheelchair they needed arrived from the manufacturer at about 10 a.m., which set in motion a frenzy of construction and programming. Working against the clock, the team built and coded furiously through the night to have something ready for Steve. He arrived at Building 92 on

Tuesday morning not knowing what to expect. Thanks to the team's all-night efforts, they were able to give him a glimpse of what was possible. Matthew promised that when Steve returned that afternoon he would be able to drive his wheelchair with his own eyes. And he did.

"The breaking news is the progress the wheelchair team has had," Matthew wrote that evening just before midnight, giddy with excitement. "The eye gaze UI has been created and we have been able to successfully pilot our little robot – Awesome! Steve Gleason today drove me around in the wheelchair with his eyes."

By Wednesday, the final day of the hackathon, in the giant white tents containing hundreds of competing projects opened on the soccer fields at Microsoft, Team Gleason had grown into a cohesive unit with a workable wheelchair that could be driven with anyone's eyes. They later would recall that making low-cost technology accessible to those most in need was a galvanizing force. And so was winning. The entire team was there pitching their creation to anyone who would listen.

"We campaigned like crazy to get people to vote for us. 'This is important, something we should be doing,'" Jon Campbell said.

Microsoft employees and executives stopped by the booth in droves. They were asked to vote and to pass along to others what they were experiencing.

Vidya Srinivasan and others on the team started tracking votes and discovered the top hacks had 180 or more votes. The company's senior leadership team would only review the top five hacks, so they got busy evangelizing. To their surprise Team Gleason hit 240 votes in early tallies, but the campaign would need to continue through August 8.

On the eve of the final count, Matthew sent a note to the entire team. Of the 3,057 total hacks, Team Gleason was sure to be one of three winners and was winning the overall vote.

"ALS shuts down the body but leaves the mind and senses untouched," he concluded. "Imagine not being able to move anything but your eyes and being reliant on someone for everything you do. Technology can be a cure and we have the ability to help an estimated 400,000 sufferers of ALS worldwide and many hundreds of thousands with similar conditions realize their independence."

Driving Ahead

The following day, they had clinched it. Not only did the Gleason hack win, but it coincided with the launch of another campaign that went viral – the ALS Ice Bucket Challenge. To generate attention for ALS research, people around the globe began dumping ice cold water on their heads and then sharing the video online. Satya Nadella decided to take the challenge and urged Larry

Page of Google and Jeff Bezos of Amazon to do the same. Matthew and Jon, with the entire team surrounding them, lifted a large bucket of ice water over their CEO's head and happily doused him with freezing cold water.

A few days later, Matthew lay near death on a mountain in the backwoods of Canada. A serious cyclist, he had scheduled himself for a cyclocross vacation right after the hackathon and had hit a jump wrong, falling to the ground and fracturing his neck and thoracic spine. For two days he was paralyzed and in intensive care. Unable to move, he thought for hours about the eye-control wheelchair hack and what it might now mean for him. Matthew and his wife were alone in the hospital and scared. They called Jenny Lay-Flurrie, and without hesitation she and her husband jumped in the car and drove eight hours north to help out. Fortunately, Matthew was soon released after he recovered, to the relief of the team.

After the hoopla of the hackathon, there were no cash prizes or big trophies handed out to the winners. No long sabbaticals or VIP campus parking spots. What the winners got was an hour with Satya. They could use the time with the CEO any way they wanted, and it was scheduled for November.

The goal of hackers everywhere is to see their inventions have impact, which usually means seeing it incorporated in a new or existing mass market product.

But this hack was still a long way from being market-ready.

"They made the tech kind of work," Jay Beavers, the longtime Microsoft engineer, would later recall. "They got proof of concepts working and then the hackathon ended. No one (on the technical arm of the team) was sure what to do after that."

But while the work paused on the technology solution, the effort to make sure that this project moved from duct tape to reality continued. Steve had gained many fans at the company, having met with leaders such as Qi Lu and even Bill Gates. Jenny and others spent hours meeting with folks across the business, on a mission to make sure that this hack project got the engineering resources it needed.

But then came turmoil, as so often happens in a large corporation. By October there was a re-org that impacted the hack team. A few of the key players were suddenly gone. To make matters worse, Microsoft eliminated its robotics group, which could have played a key role in building the next version of the wheelchair.

The breakthrough came with Jay, who loved robotics and the integration of hardware and software. He saw the wheelchair as a perfect application of his passions. And if the darkest hour is just before dawn, hints of a sunrise appeared on the horizon. As if somehow meant to be, things began to fall into place. Peter Lee, the celebrated computer scientist in Microsoft Research, formed a new team he called New Experiences in Technology, or NExT, to apply exciting research ideas to concrete commercial product development.

Jay decided to pitch Peter on incorporating eye gaze as one of the first NExT projects. Unfortunately, the hackathon had ended and Jay had to go find what remained of the wheelchair used in the hack. He located it in a closet deep in the bowels of Microsoft, tucked away in a moving box. Three days later he drove the wheelchair into Peter's office.

"This is the kind of project you want to do," he told Peter. "If you give me five people, I can make this happen."

Peter jumped at the opportunity, naming Rico Malvar, a top distinguished scientist, as executive sponsor of the project. Everyone cheered.

Looking ahead to the Satya meeting in November, the team decided to go big or go home. They would ask for eye tracking functionality to become a standard part of the Windows operating system. And they believed Microsoft should build and market a new Windows – device: an eye gaze wheelchair. They set to work putting some structure around the hack. They would incubate the wheelchair within NExT.

As the weeks flew by, the team felt the high stakes of a meeting with the CEO. They labored over a pitch deck for Satya and prepared for the kind of adversarial meeting that had been Microsoft legend. They anticipated arguments like, "This is too expensive; the market is too small; this will never work."

Finally, when they were invited into Satya's office for the pitch, the team was loaded for bear. They had rehearsed and re-rehearsed their arguments. A mix of anxiety and excitement spread through the team. But instead of precision questioning from the CEO, they were asked a question they had not expected.

"You're starting a team to do this, right?" Satya asked Rico.

Instead of begging for the opportunity to bring the eye gaze wheelchair to market, they shifted immediately to how they were going to do it.

There had been rumors since Satya was named CEO that his son had cerebral palsy and relied on a wheelchair, but no one knew for sure.

"My wife would cry if she saw this," Satya told the stunned room as he sat in the wheelchair and moved it around the room with his eyes.

They left the room committed to take the essence of the original hack – independent mobility and natural communications – into the new group. Steve Gleason and other PALS (People with ALS) remained part of the development team.

The Tech Behind the Hack

In order to understand the project in its entirety, what follows is a time-compressed summary of the technology that evolved from the nuts-and-bolts the hackers cobbled together during their intensive 48-hour sprint and weeks prepping in advance.

Over time, Jay Beavers, the bearded and joyful senior programmer and researcher with experience in robotics, caught wind of the hack from fellow software developer and engineer, Ashley Feniello. Jay and his newly formed team jumped into the hack, and together they painted a roadmap for where the project was going.

Figure: A roadmap for the eye control wheelchair:

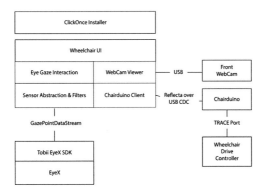

The engineers began with ClickOnce, a technology that allows users to install a Windows desktop application as simply as clicking on a link in a web page. The browser downloads a little file which Windows then knows how to use to download, run and update the application.

Next, the application and user interface for the wheelchair is written in Windows Presentation Foundation (WPF), a technology for making Windows desktop applications. This is an "older" type of application that can't be distributed through the Windows Store, which preceded Windows 10. It's not impossible to talk to interface from a Windows Store app, but it's a lot more complicated so they used a Windows desktop app to keep things simpler. The wheelchair UI is very simple -- translucent directional movement buttons overlaying a full screen webcam video of the forward-facing camera. They used the background video to try and help the Surface disappear.

Steve Gleason's future wheelchair was designed to get simple data from the eye gaze sensor — a stream of X/Y coordinates telling the system where he would be looking. The hackers turned those into interactions, or clicks, when the gaze paused for a period of time.

To help distinguish when Steve was looking at an element like a button, textbox or tab, they used WPF's VisualStateManager to change its look after a short delay like 250 milliseconds and then perform the activation, or click, after an additional longer delay, 250 to 500 milliseconds. For some buttons, such as 'show settings' or 'exit app,' they set a longer activation delay to prevent accidental clicks.

The code they wrote went something like this:

```
VisualStateManager.GoToState(hitTarget, "Click",
true);
    We activate elements using WPF's AutomationPeers,
for example
    var button = hitTarget as Button;   as Button;
        if (button != null)
        {
          var peer = new
ButtonAutomationPeer(button);
          var provider =
(IInvokeProvider)peer.GetPattern(PatternInterface.Invo
ke);
            provider.Invoke();
            return;
        }
        vartoggleButton = hitTarget as ToggleButton;
        if (toggleButton != null)
        {
          var peer = new
ToggleButtonAutomationPeer(toggleButton);
          var provider =
(IToggleProvider)peer.GetPattern(PatternInterface.Tog
gle);
            provider.Toggle();
```

```
    return;
}
var textbox = hitTarget as TextBox;TextBox;
if (textbox != null)
{
    textbox.Focus();
    return;
}
vartabItem = hitTarget as TabItem;
if (tabItem != null)
{
    tabItem.IsSelected = true;
    return;
```

A critical phase of the hack was safety. It was essential to know when the wheelchair driver was no longer looking at the screen or had closed his or her eyes. They designed the system to check incoming data for idle timeouts, including no data coming in because no eyes were detected.

And they wanted their system to work with many different types of sensors. This allows for comparison shopping to get the best price and performance. The market for eye gaze sensors has been disrupted over the past few years so prices dropped from about ~$2,000 to, eventually, under $150. The latest and greatest was the goal.

They had to implement multiple filters to test different algorithms for dealing with the high noise emanating from the sensors. The filter most commonly used was called the GainFilter with a simple formula they expressed in this way:

```
double distance = Math.Sqrt((( _filteredX - measuredX) * (_filteredX - measuredX)) + ((_filteredY - measuredY) * (_filteredY - measuredY)));
        if (distance > SaccadeDistance)
        {
          _filteredX = measuredX;
          _filteredY = measuredY;
        }
        else
        {
          _filteredX = _filteredX + (Gain * (measuredX - _filteredX));
          _filteredY = _filteredY + (Gain * (measuredY - _filteredY));
        }
```

SaccadeDistance is used to disengage the filtering when a rapid eye movement (or saccade) occurs. The SaccadeDistance should be set to around 'one half of the average movement distance between targets' so that when a person moves her gaze over halfway between one target to another, the filter 'releases' the gaze to allow it to quickly travel. In our testing, we generally find a value of 0.07 to be a good initial SaccadeDistance for our applications. The Gain should be set based on the expected noise in the system. In our testing for a

Tobii EyeX Gain for indoor settings with low external IR interference should be set between 0.04 for a precise user to 0.07 for an imprecise user. As the noise increases, either due to poor eye modeling conditions (thick glasses, dirty glasses, bifocals, older eyes, astigmatism, etc.) or external interference (e.g. halogen lights or sunlight), the Gain should be increased.

They translated the Tobii EyeX SDK into their own device-agnostic GazeEvent data format.

They do not leverage the WPF integration libraries that Tobii provides as part of the EyeX SDK as this would tightly couple our system to just the Tobii EyeX.

Finally, the Chairduino control board communicates to the app via a protocol called Reflecta, basically a remote procedure call (or RPC) client for microcontrollers like the Arduino. The Chairduino client is essentially just ReflectaClient plus a wheelchair-specific interface named 'whel1'. 'whel1' defines a remote function called ListPins which tells the wheelchair app which digital output pin refers to which drive command on the TRACE port.

The team ended up switching from Firmata to Reflecta to include additional safety intelligence to the Chairduino, such as keepalive packets, auto-stop if the Chairduino loses connection to the wheelchair app, and a watchdog timer to detect firmware crashes in the Chairduino itself.

The Chairduino Firmware supported functions like setting status LEDs that confirm communications between the Chairduino and the wheelchair, the Chairduino and the wheelchair app, and the heartbeat of the firmware itself. The system includes a watchdog timer that resets the Chairduino if it crashes or hangs and a triple-blink on startup so they could visually detect a crash or restart.

- - -

A lot of people were talking about inclusive design, the notion of designing products that work for everyone, not just people with disabilities and not just the broad consumer market – everyone.

The team was doing this for real. And they were doing it for everyone.

By then Jon Campbell was working on Microsoft's virtual assistant, Cortana, and was volunteering his time to the wheelchair project, as if it were still a hack. He'd attend meetings where Jay Beavers and Rico Malvar would tell the group they still had headcount. Finally, someone looked at Jon and said, "Dude, they are talking to you."

"I worried about going into research." Jon said. "I always told myself it was important to work for a group that makes money. Is this team going to be a fad? Is this a six-month effort?"

"I had to work on him," Jay said. "We were going to make this happen and get it into the market. We now had the resources and the executive support to do it."

Jon went through a rigorous interview process and was convinced. But the next steps were more complex, and beyond Microsoft's traditional experience. A wheelchair is considered a medical device, and so safety is paramount.

According to the U.S. Food and Drug Administration (FDA), a powered wheelchair is a Class II medical device, which translates into "greater regulatory controls to provide reasonable assurance of safety and effectiveness." Nevertheless, states like Louisiana, home to Steve Gleason, were experimenting with new right-to-try laws that gave terminally ill patients much faster access to experimental drugs and devices.

By 2015 they finally had a solid prototype of what today is called Eye Control in Windows. A wheelchair could now be driven by a person with his or her eyes while sitting in the wheelchair – not an autonomous wheelchair. And so Jon and Jay flew down to New Orleans to demonstrate the latest version for Team Gleason.

Everyone was excited to see how it had evolved, and one afternoon Jon sat in amazement as he watched Steve chase his 4-year-old son around the room – playing together as had been Steve's dream.

"That was a powerful moment for me," Jon remembered.

When it was time to return to Redmond, Jay and Jon started to pack up the prototype wheelchair. They had gotten invaluable feedback and it was time to return to campus for more refinements. But through his synthetic voice translator, Steve told them "no.

"You're not taking this. Make a new one and refine that. This remains here with me."

Message delivered.

Bug Bash

It's mid-January 2018 in a nondescript Microsoft conference room, where a dozen or so developers have gathered for the latest "bug bash." The team is on a mission to run Eye Control in Windows through the gamut of tests to identify errors, problems and other bugs. They are just days away from an important review in Houston, Texas, not only with Steve but a number of wheelchair manufacturers and marketers who gather for an annual abilities conference.

Jake Cohen is a program manager on the Windows Interaction Platform team, and is one of the people responsible for designing and integrating Eye Control into Windows 10.

"This is all about making the next Windows Update awesome for every person using eye gaze," Jake announces. "Let's log as many bugs as we can."

Working alongside him are Jay and Jon, as well as principal development lead Harish Kulkarni, principal user experience lead Ann Paradiso, Irina Spiridonova (the team calls her "the goddess of testing"), and developer Austin Hodges.

"We didn't start as accessibility pros," according to Jon. "We sought these team members out. We're breaking the mold of traditional paradigms."

"We have the best customers in the world," says Jake.

The team accompanies our customers, people living with ALS, to their homes, spends time with family and joins doctor appointments, all to see the successes and opportunities those living with ALS see day in, day out.

"Every single feature released in Windows comes directly from customer feedback," according to Jon. "The next release is always variable because it requires so many teams to fall into place."

In this latest release, the team is launching scrolling with the eye, and direct right and left click - the two features most desired by users with ALS.

The team has just released a set of APIs so that any developer out there can engineer his or her own ideas.

"APIs are an opportunity for us to lead," says Harish. "If we do this right, we create an underlying API that's incorporated into common controls. And with that, any developer or org using common controls lights up."

Still, the backlog is long, and other issues challenge customer satisfaction with the impending release. Some eye colors are harder to track, and something as simple as adjusting your eyeglasses can disrupt calibration between the pupil and the eye tracking device. There remains a lot of work to do, and there is competition. Apple, Google and Facebook have acquired eye tracking companies. Empower someone with ALS to easily and effectively use technology, and you may also resolve barriers for Virtual and Augment Reality optics and telepresence for everyone. We're already seeing this with animated emojis from Apple and facial recognition from Snapchat and Instagram. Eye tracking will take once lifeless avatars to the last mile.

"What is the equivalent of eye contact in telepresence settings?" Jon asks.

Meetings for collaboration and communications are crucial scenarios for any productivity software company. In a meeting, eye contact is essential. With technology like Eye Control, the human connection, social interaction, in a virtual setting can be dramatically improved. The original principles of universal design were making technology more accessible and advancing competitive product innovations.

The bug bash continued through the day and ended with a sense of urgency.

Houston, we have a solution

The lobby of the Marriott Marquis in downtown Houston, Texas, is noticeably crowded with wheelchair traffic. It's late January 2018, three and a half years after the original Eye Gaze hack, and several members of the team have flown from Redmond to meet Team Gleason and a dozen or so assistive technology professionals, including wheelchair manufacturers and distributors, to provide an update on eye control in Windows. Upon entering elevators, guests are greeted by enormous photographs of astronauts and other space motifs, a relic of Houston's NASA heritage. But today it's more a reflection of the moon-shot people with disabilities expect from assistive technologies.

They've gathered to attend the annual leadership conference convened by Numotion, a leading provider of wheelchairs. Also attending are representatives from Stealth Products, a manufacturer of wheelchair accessories like joysticks and head and neck positioning systems as well as Quantum, a manufacturer of wheelchair power bases and driving controls.

A few minutes after 11 am Steve Gleason can be heard approaching the conference room from out in the hallway. He's talking to his team over the loudspeaker that projects words typed with his eyes onto a Surface, and out through a synthetic version of his own voice, which he "banked," or recorded, years ago before ALS took his vocal chords. He rolls into the room where Microsoft, Numotion, Quantum and Stealth await along with Jay Smith, founder of the music technology company Livid who also lives with ALS and is experimenting with his own version of Microsoft's eye control technologies. Gleason has grown a handsome, grey beard that is flecked with dark whiskers. He's strikingly handsome with the same athletic eyes trained onto his Surface screen. He's ringed by his caretakers and Team Gleason advocates, also athletic men in t-shirts that say, "Rebirth," "No White Flags," and "37," his New Orleans Saints jersey number.

"Howdy everyone," he says, positioning himself just across from Blair Casey, the assistant executive director of Team Gleason who this day is resplendent in a sharp suit and a 60s era tie.

"I wanted to have a direct view of Blair's tiny tie."

The room erupts in laughter, releasing that little bit of tension around the table. Jay Beavers kicks off the discussion with a reflection on the state of the technology.

"Everyone saw Steve driving around the room so that's a good place to start," Beavers says. "He's got a Surface with a Tobii eye-tracking bar. The red dot on the screen is where Steve's eye is looking. Steve is working as a tester of our eye control in Windows. Is it working well, what do you think?"

"It's amazing," Steve confirms. "I am using the eye drive with multiple speeds."

Independent of the Microsoft effort with Steve Gleason, Jay Smith, a technologist and entrepreneur who is joining the meeting, has built his own version of the Microsoft eye control.

"We've got an engineer and an athlete, so they are able to process everything very quickly."

Jay stands up and walks over to a vacant wheelchair with the eye control system pre-installed. He sits down, calibrates the eye control and begins to drive it around the room, forward and backward, making circles and narrating his activities.

"Like a product about to go to market, it is totally unremarkable in that you plug it in and it just works," Jay says after his demo.

A representative from Stealth Products chimes in

that they want to see the eye drive interface work with multiple wheelchairs and multiple devices.

"I really don't see this work as device-limited. Am I wrong?" Gleason asks.

Jay Beavers notes the device needs to be lightweight but can work on a range of tablets. And he reports that the Universal Serial Bus (USB) compliance committee has now agreed to have eye control as a Windows feature for head tracking standard, and it's now included in Windows 10.

"Excellent news," Steve says

The conversation switches to the heart of the matter. How do we get this technology to those who need it most, which means government approval and Medicare and Medicaid-funded individuals? There is discussion of FDA approvals, government relations and the need for Medicare and Medicaid codes that patients can use for reimbursement.

"We don't want to build a rocket no one can ride into space," one developer said.

Collectively, everyone agrees the more this technology gets out there in the market, the more and quicker feedback, learning and improvement can happen.

"We can accomplish anything with software that we set our minds to, but the downside is that we can do everything we set our minds to and never be finished," Beavers said.

The eye-gaze wheelchair is undergoing a clinical trial at Swedish Medical Center, the largest nonprofit health provider in the Seattle area. And FDA approval, which will examine issues like wireless co-existence, sunlight tolerance and temperature thresholds, is on the horizon. And there is the infamous test in which the wheelchair is dropped 12,000 times from a one-foot height to test its durability.

"There are 2,000 people out there waiting for this," Clare Durrett, Team Gleason's associate executive director reminds everyone. The team from Stealth points out that when you add the spinal community and those with "locked-in" syndrome the numbers in need are quite large.

Steve Gleason and Jay Smith muse about issues ranging from whether the wheelchair should have a camera for reverse – not necessary – and the ability to look away briefly from the eye tracker. In the current build, the wheelchair stops moving if the driver looks away. Gleason tells the engineers he prefers to try something before ruling it out. They all agree that collision avoidance technology, like that in an automobile, will be an important feature to add down the road. Training patients is something also on the roadmap. The priority now is to get the best product possible tested and on the market.

As the morning hours fade into late afternoon, Jay Beavers encouraged the group to set a check-in meeting and to avoid retreating to their individual camps. He outlined roles and responsibilities across testing, investment, partnerships and marketing.

Both Steve Gleason and Jay Smith announce they want to remain very involved, "hands on, so to speak," they chuckle.

"We've made progress but I am worried about losing momentum as we go into the regulation phase. Let's leave with commitments," Gleason implores.

They agree to bi-weekly meetings and keeping things moving ahead as a unit. Their goal is 12-14 months to market.

Not long after the Houston gathering, Steve Gleason posted an essay to LinkedIn reflecting on his experience.

"Technology can be a cure. I will put it bluntly. Instead of being a productive member of society and a pretty bad ass father, without technology, I would be dead. I would call that a cure. Using just my eyes, I am able to type, speak, and navigate my Microsoft Surface Pro tablet – no different than any of you ordinary humans. I can text, tweet, host webinars, even stream any song. I can drive my wheelchair with my eyes. This is entirely liberating."

All this from one simple email that missed its target when it first went out.

Microsoft 2014 Hackathon EyeGaze team - Row 1: Dave Gaines, Vidya Srinivasan, Jenny Lay-Flurrie, Matthew Mack (orange shirt), Jon Campbell, Erin Beneteau (on stairs) Row 2: Bryan Howell, Bruce Bracken (camera & tripod), Genevieve Alvarez

Photo by Scott Eklund, Red Box Pictures.

The Ability Eye Gaze team, which won the Grand Prize in Microsoft's first-ever companywide Hackathon, dumps a bucket of ice water on Microsoft CEO Satya Nadella as part of an effort to raise awareness around amyotrophic lateral sclerosis (ALS).

Photo by Scott Eklund, Red Box Pictures.

"Until there is a cure for ALS, technology is a cure." Steve Gleason, pictured here in August 2015, is a former pro football player who is living with amyotrophic lateral sclerosis. The eye gaze project united two dozen researchers, engineers, designers, program managers and media pros from across Microsoft, and was one of the 3,000-plus teams that participated in Microsoft's first-ever global Hackathon. The team aimed to use Surface 3, Kinect and other Microsoft technologies to create a wheelchair that Gleason could drive with his eyes.

Photo by Brian Smale

OneNote Hackathon finalists OneNote team in Vancouver, Canada (left to right) Mark Flores, Alex Pereira, Sebastian Greaves, Pelle Nielsen, Scott Leong, Dominik Messinger, Reza Jooyandeh, Ken Wong on August 11, 2015.

Photo by Scott Eklund/Red Box Pictures

In 2014, the Hackathon replaced Microsoft's annual company meeting. It's part of One Week, now held the last week in July at Microsoft's Redmond campus, with corresponding events worldwide. Pictured here is the opening ceremony in 2015. The 2018 Hackathon is expected to draw 5,000-7,000 participants in Redmond, 18,000-20,000 around the world.

Photo by Scott Eklund/Red Box Pictures

–
An Interlude

POETIC

In 2008, Philip Schultz won the Pulitzer Prize for poetry. The Washington Post once described Schultz's language as "plain-spoken and the works are replete with insights and nuggets of wisdom."

Around 80 people in history had ever won a Pulitzer for poetry. As far as we know, only one with dyslexia has ever won — Philip Schultz.

Schultz writes movingly of his life — and his journey from the table for "dummies" in grade school to becoming a world class poet — in a slim volume published in 2011 entitled, *My Dyslexia* (W.W. Norton & Company). He was 58 when he finally figured it out, and only after his son was diagnosed with dyslexia.

"Suddenly everyone I knew seemed to have either

suffered from some kind of learning disability or knew someone who did. And I mean everyone."

Dyslexia affects one in five people. As defined by the International Dyslexia Association:

"Dyslexia is a specific learning disability that is neurobiological in origin. It is characterized by difficulties with accurate and/or fluent word recognition and by poor spelling and decoding abilities. These difficulties typically result from a deficit in the phonological component of language that is often unexpected in relation to other cognitive abilities and the provision of effective classroom instruction. Secondary consequences may include problems in reading comprehension and reduced reading experience that can impede growth of vocabulary and background knowledge."

Albert Einstein, also dyslexic, said he wasn't the smartest person he knew – others had higher IQ – but he was the most creative.

"In a sense, dyslexics are conditioned by their environments to blame only themselves for their learning difficulties. Interview any dyslexic and you'll soon discover a world of blame, guilt and shame," Schultz wrote.

–
Part II
LEARNING TOOLS

"If you design things for the greatest accessibility —
Learning Tools is like that — it makes everything
accessible to all, and why wouldn't we want that?"

-Fourth grade teacher

An Education

It takes a lot to get under Mike Tholfsen's skin. But he'd had it.

"People just didn't care," he told me one day in a conference room near his desk in Building 6. Years had passed, but he retold the story like it was yesterday. "In 2010, I went cold turkey on education, joined a new group and closed that chapter of my life."

The chapter Mike thought he had closed was his almost obsessive pursuit of designing software in support of education. He firmly believed that OneNote, the Microsoft Office product that helps users organize, collaborate and share rich content, could transform education for teachers, students, administrators and parents. His dad had been a teacher and professor, his mom a librarian. But that wasn't the only reason he cared so much. He just remembers often sitting in a meeting and imagining OneNote as a helpful tool for curriculum, pedagogy, teacher feedback, collaboration. An entire classroom, a whole school for that matter, could be digitized and made more accessible with OneNote. From then on, that's all he could think about.

He had joined the company back in its golden age, 1995, and had worked on version 1.0 of Outlook with Brian MacDonald, a longtime Microsoft executive with a long list of product hits to his credit. By 2004, Mike was an engineer and a lead for a new product that was looking to get traction, OneNote. The more he thought about its application to education, the more time he spent on thinking about OneNote and schools. He talked with any and every educator he could find. In classrooms, on the street, across the backyard fence. It didn't matter.

"I was the crazy OneNote education person," he laughed.

But the joke was on him. No one at Microsoft was

receptive to his ideas. So, he left the OneNote team and moved over to do "cloud stuff," like Office 365. He tried to forget about schools and software. Time went by, O365 caught fire with consumer and business users. Life was good.

Then, in August 2013, he noticed an email in his inbox from Chris Pratley, the leader who first incubated OneNote. Chris had added Mike to an email thread with Jonathan Grudin, a researcher within Microsoft Research who focuses on natural human-computer interaction. Jonathan had written a white paper about how OneNote might be a transformational tool for education. Mind you, Mike was no longer working for OneNote, but he bundled up all of his files, notes and prototypes from work with teachers, and together with Jonathan conceived the idea of Class Notebooks. They pitched their idea to Chris who loved it, but had no resources to offer. So, they cobbled together enough funding for a dev team in China that could help, and soon they had what he remembers today as "a crappy OneNote app." They put it out to teachers for feedback, fixed the bugs and continued to add new features.

"The passion got reignited," Mike said. But his next generation learning approach was going nowhere fast.

Things were about to get interesting. Two Microsofties from different parts of the company were soon to play an outsized role in advancing Mike's dream. A former management consultant named Jeff

Petty joined Microsoft that year as a strategist with a focus on accessible technology. And a leader named Eran Megiddo took over the fledgling OneNote product.

Jeff, the son of a Cold War engineer who also led development of the optics for Hubble Space Telescope in the 1990s, was not satisfied with the role of simply ensuring Microsoft technology was in compliance with the government's disability regulations. He wanted more. Like his dad, he was motivated to do great work and make a difference. Just a few months into the job he rallied a team to join the 2014 hackathon with an ability project they called "Audio Lens," a hack of the Windows audio channel that would personalize and thereby improve sound for those experiencing hearing loss. The goal was to optimize and tailor the voice and music experience, but the hack fell short, and Team Gleason went on to win. The loss stung, but Jeff vowed to learn from it. Meanwhile, Eran, a multilingual Israeli known as a force of nature for his abundant ideas and lack of patience with bureaucracy, took over as the head of OneNote. He had a startup mentality and believed fervently in technology as a vital tool for education and accessibility.

Getting more people on the bus

That fall, not long after the success of Microsoft's first hackathon, Mike pitched Eran on OneNote for education and suddenly found himself back on the team. He also regained leadership over his old Class

Notebooks application, returned to his previous obsession of talking to every educator he could find and, with just one engineer, iterated, revised and improved the product little by little.

But he still wasn't satisfied. The more he observed classrooms and talked with educators, the more he came to realize the profound challenge of children with learning disabilities. In fact, 1 in 5 students has a learning disability. Dyslexia, a general term for disorders that make it difficult for kids to read and learn, can be debilitating for the student and can slow the overall progress of a classroom as teachers spend more time on individual remediation.

Mike knew he needed to get more people on the bus. He discovered a man in France, Dennis Mason, who was using OneNote to help students with dyslexia. To get the conversation bubbling within Microsoft, Mike funded a video about Mason to highlight how OneNote was being hacked to help students with dyslexia improve reading and comprehension. He then began to take engineers with him to education conferences in hopes of building understanding and empathy.

Mike, Scott Leong and Valentin Dobre love to tell the story of how they were standing at the Microsoft booth at South-by-Southwest Education, a popular conference in Austin, Texas, when a woman wandered by asking for directions. They suggested some directions, and then told the woman what they were up to. She couldn't believe it. Just that morning, her dyslexic daughter had crawled under a dining table, got into the fetal position and cried because she couldn't read well enough to keep up in class. The mom joined her under the table, also crying. Mike gave Scott and Valentin a side-long glance. He could see they were inspired.

Meanwhile, across the Microsoft campus, Jeff Petty was investigating multiple options for advancing the company's strategy for making its technology more accessible for people with disabilities. He was increasingly convinced that applications that could help with reading and writing could be an area of differentiation for the company. He was so convinced that he pitched the Office and Windows teams about purchasing an independent company that offered learning and literacy solutions, but the direction he got was to go build it rather than buy it. That's when he went to find Eran Megiddo and Mike Tholfsen.

"Mike believed OneNote was the killer app for education," Jeff recalled.

Eran tasked a group of young Microsoft developers in Vancouver, British Columbia, 120 miles north of the Microsoft campus, to help advance OneNote for education with a particular focus on students with learning disabilities. Mike and Jeff were now joined at the hip, and decided to take a road trip together to Vancouver to meet their new teammates. As they approached the Canadian border, Mike and Jeff agreed to aim their aspirations for a public display that coming summer at Microsoft's second annual hackathon.

They pulled into a parking spot in Vancouver's now trendy Yaletown, a downtown community that traces its roots back to the gold rush in the 1850s, a neighborhood where Microsoft had what the development team leader Scott Leong described as a "dark and dingy" office. There they found Reza Jooyandeh, an Iranian, Sebastian Greaves, a Brit, Pelle Nielsen, a Dane, Dominik Messinger, a German, and Valentin Dobre, a Romanian, were all enthusiastic about coming to the United States to join Microsoft. Scott, a native Vancouverite, was thrilled to have a serious software development office with a startup feel right in his backyard. He knew from Eran that their new CEO, Satya Nadella, felt the company needed to jumpstart new development projects. In the words of composer, lyricist and playwright Lin-Manuel Miranda in *Hamilton*, they wanted to show that "immigrants, we get the job done."

Collectively, they all had a lot to learn. Jeff had found a plethora of resources hiding right under their noses at Microsoft in the form of experts and untapped innovations. Among the most influential experts he found were reading specialists Greg Hitchcock, Kevin Larson and Rob McKaughan. Greg's Advanced Reading Technologies team, loaded with expertise in type design, reading psychology, human vision science and engineering, had the broad mission of improving on-screen reading. After all, products like Word and browsing are all about reading and writing. For a long time, Greg, Kevin and Rob were focused on trying to get on-screen reading to reach parity with printed material, and improving on-screen fonts, rendering and typographic layout. The reading experts were able to replace some of the early enthusiasm for so-called dyslexic fonts with more research-based approaches like text spacing.

In their first meeting, the reading team showed Jeff a presentation. He still remembers the awakening that came with the very first slide.

"Our 500-year-old conventions for punctuation and layout are serviceable but not optimal. We can do better. For instance, spoken sentences contain more information than written sentences. In addition to the words themselves, we speak information about pronunciation, syntax, stress and emotion. Dyslexic and low vision readers have different visual needs than other readers. This talk will show a variety of reading comprehension enhancements that we can implement today."

From then on, the reading team was an integral part of the development team.

The hackathon cometh

The summer hackathon was approaching fast. Just as Team Gleason had prepared for the first companywide hack the year before, Mike, Jeff, the reading team and the Vancouver developers began in earnest to think about entering OneNote Learning Tools in the second Microsoft hackathon – not just entering, but winning it.

It was about this time that Jeff discovered a field research study conducted by Mira Shah, who had been with the company just over a year and had participated in the previous year's winning eye gaze hack team. As a result of that work, she had gotten more involved in accessibility issues at the company and met with lots of people in that community – people with disabilities, parents, special education teachers. One part of her report included the story of a specialist that would enter a classroom to help one student on what's called an Individual Education Program (IEP), a written plan for helping to meet that child's needs (in this case the child was dyslexic). The specialist realized that in any classroom there were many more students who needed help but had not yet been screened. The specialist would bring assistive technology apps – first one and then the other – into the classroom to find what helped most. It was inefficient and inequitable.

The experience she brought to the hack team, though, was far richer than anyone could know. Prior to Microsoft, Mira had been trained as a speech language pathologist and continued to maintain her license to practice. In college she had wanted to become an educator who worked with people who are deaf. Upon learning her school offered a speech pathology degree, not a deaf education credential, she pursued that. She fell in love with language and how people acquire it. She went on to graduate school for speech pathology and landed at Shriners Hospitals for Children in Portland, Oregon. There, she designed wheelchairs integrated with other assistive technologies for children as young as 24 months. She was the only language therapist on staff. In time she designed curriculum for kids who use assistive technology and built a popular app. That's when she was hired in 2013 by Microsoft.

"I always wanted to leverage my experience with assistive technology consumers," she said. "I wanted to impact the direction of assistive technology and to move into consumer platforms."

Mira became the Learning Tools' subject matter expert, analyzing the competition and helping with the screen reader, dictation and syllabification.

Meanwhile, Jeff continued to pay attention to what the reading team was telling him, and he began to discover services, tools and programs scattered around Microsoft that could make a real difference if applied to a unified OneNote learning tool. Chris Quirk in Microsoft Research had a natural language and grammar parsing service, which meant they didn't have to invent new ways of creating syllabification – breaking words down into syllables to improve readability. He turned to Bing speech services in order to turn speech to text and enable the computer to read out loud – all important functionality for people with dyslexia.

"We didn't have to build the magic, we just have to plug it in, which was different from the previous year's Audio Lens hack."

The team found that curiosity, asking questions and being open got them up out of their individual org charts to see more of the playing field.

"Reading knows no organizational charts," Jeff said.

They were now in full hack mode. There were weekly calls between Redmond and Vancouver. Jenny Lay-Flurrie, still the chair of the disability employee group, had grown the ability hackathon category for 2015 from 10 to 70 projects. She sensed something important was happening with what the team was now calling OneNote Learning Tools.

Several members of the Vancouver development team were given permission to travel to Redmond for the final days of the hack. Dominik, Sebastian and Pelle piled into Seb's car and headed south. At the border crossing, Dom forgot to execute a document and found himself face-to-face with the grumpiest border guard he'd ever met. He was told his mistake could be a felony. To ease the tension he piped up, "A Dane, a German and a Brit cross the border together, sounds like a bad joke, huh?" The border guard failed to see the humor, but after a long delay and some explaining they were told to get on their way. They checked in to a hotel in Seattle's South Lake Union neighborhood and hustled to the Microsoft campus to start building the immersive reader. Jeff and the team were the first ones into the tent that year.

At the booth they set up two whiteboards — one for conceptualizing what needed to get done and the other for tracking the work and status. They had it down to a science. Every two to three hours they would have a checkpoint, but they had to give the developers enough time to actually make progress. Sprint, meet, sprint, meet, repeat.

They had their work cut out for them, and they shared responsibilities across all of the features. Their goal was to increase reading and comprehension scores for people diagnosed with learning disabilities. The arsenal of features they sought to accomplish was long and complicated: read aloud, dictation or speech-to-text, identification of English's parts of speech, syllabification, breaking sentences into smaller parts (also known as comprehension mode), and font spacing and short lines. Greg Hitchcock, inventor of favorite fonts like Times New Roman and Arial, created a new font that solved a confounding reading barrier for dyslexics called visual crowding.

They started implementing the various services,

working on the User Interface (UI), rewriting code and fixing things.

"Things were crashing," Pelle recalled.

It was all hands on deck. Some of the services were running slow. They used tech that was not perfect. They were rewriting scroll bars and drop down menus – very basic stuff. Dom suddenly had to write a toggle button. They were throwing files on disks since some of the services were experimental and not yet in the cloud.

At one point they downloaded a bunch of words broken into syllables and the non-native English speakers learned a lot of new swear words.

It was the most exhausting and thrilling work any of them could remember. The hours were very long. At one point Dom had to go outside the tent, find a park bench and lie down to sleep. Seb said he was so knackered he fell into bed well past midnight each night.

With 10 minutes left on the clock, OneNote Learning Tools just broke. Frantically they diagnosed the problem and fixed it with seconds to spare.

As they reached exhaustion, that's when the campaigning began. Jeff kept his eye out for any executive that happened to be wandering by. The team members tasked with marketing, led by Dan Hubbell, had created a compelling video and they went to the hackathon's science fair, a public forum, where they pitched their work like carnival barkers. They understood that landing the story of the product was as important as the product itself.

Drained, they all returned to their offices. And for days they sat at their computers staring at the hackathon website known as the Garage. One afternoon Jeff got a call asking if the Redmond and Vancouver teams could be photographed so he suspected something good had happened, but he didn't know. Then, without further warning, the Garage website announced that the OneNote Learning Tools had won the grand prize.

"We got a block of wood, a red jacket and a shit ton of work," Jeff chuckled.

Moving forward and changing lives

Jeff and Mike fondly remember Eran's response to their victory. "Buckle up this is going to be big. What do you need? How fast can you go? And how can I help?"

Today Mike leads two teams of engineers, Class Notebook and Learning Tools. Jeff is a Windows accessibility leader charged with creating delightful Microsoft and partner experiences for people with disabilities. Like the eye gaze Wheelchair team before it, the Learning Tools team evolved from a passionate hackathon into a strategic business with a noble purpose. The tech continued to deepen and awareness about its benefits spread.

By spring of 2018, Microsoft had taken OneNote Learning Tools and built it across a dozen or more apps and platforms, including Office and the Edge browser. The Learning Tools team has built a full cloud service powered in part by Microsoft Cognitive Services, and the team has added more capabilities drawn from research and talking extensively to students and educators. The tools are available in more than 40 languages, and achieved 13 million monthly active users, an astounding feat. The team now thinks it can achieve 20 million monthly active users in the not so distant future. News of this growth coincided with word from one quarterly report showing Windows share grew an additional 6.5 points on devices under $300 – reaching its highest share in four years in this category.

Every day the team hears new stories about how their tools are helping. Not long after an education event in New York City that showcased Learning Tools, Mitra Niknam, the mother of a young boy named Andrew who had just been diagnosed with dyslexia, stopped by her local bakery. As she was checking out, she told the baker about her son's diagnosis. The baker told her he had just heard that Microsoft was doing something to help kids with dyslexia. She went home and immediately looked it up online. She couldn't believe her eyes. She called a Microsoft store in New Jersey and booked a session for her son to download the tools and learn to use them. It clicked with Andrew, and over the summer he spent hours learning to read.

"He went from basically nothing to having the world at his fingertips and being able to read. It was a miracle," Mitra said.

In Canton, Georgia, special education teacher Lauren Pittman also was desperate to improve reading outcomes for her elementary school students. Kids would come to her and say they were stupid. Lauren was one of the first teachers to try out a very early version of OneNote Learning Tools. She learned about the new OneNote Learning Tools but worried it would take forever to integrate new technology into the classroom. Instead it took just three days for the students to master the program. After introducing immersive reader she began to conduct weekly reading speed checks, wondering if it would make a difference. It did. All students benefitted. One student went from 4 words per minute went to 22 words per minute.

"I never thought in one calendar school year we could get to double digits, and he stayed there," Lauren said.

Rachel Montisano, a first grade teacher at the Ashton Elementary School in Sarasota, Florida, wrote the team to say she wanted to increase overall student achievement through heightened student engagement, active learning and student accountability. She pointed out that first grade students tend to have a wide range of reading levels with many struggling to read. They used the immersive reader to highlight words, and with text and speech they can see and hear words. They can pinch to make words bigger or smaller and no more time is wasted handing out collecting papers.

"As the year progressed, students were learning at increasing speeds, and their gains were becoming more and more apparent," Rachel said. "They took their mid-year iReady diagnostic assessment, and at the halfway point, 67 percent of my students had already made gains equivalent to one year of growth. The remaining 33 percent had started the year struggling. Of this remaining group, however, 89 percent of them were within 3-7 points of making a year's worth of growth. Some who were far below grade level now had the chance, quite possibly, to be proficient by the end of the year, given their rates of growth."

And in the Anacostia community of Washington, D.C., Merlyne Graves' classroom has become a hive of tech-assisted learning. Merlyne is a 16-year, highly-effective teacher who is constantly in search of tools that can help her students. Garfield Elementary, where she teaches fourth grade, is located in Ward 8, a low-income neighborhood. Her students face many socio-economic challenges, one of which is that they read several levels below their grade level average.

"Over the break I looked at what Microsoft was offering," Merlyne said. "Our school had switched to

Office 365 so I started watching some training videos about OneNote and Learning Tools. I thought, my god, this can revolutionize my job. I kept earning certificates and badges."

She set the goal of having a paperless office, and wrote a grant proposal, a sort of go-fund-me campaign, for the teacher-support website Donors Choose. Bingo! She raised enough to purchase five inexpensive Windows 10 tablets, and the Garfield principal became so excited he decided to help her purchase enough for the entire classroom – the hope being that she would experiment with Learning Tools and then spread the practice throughout the school.

Her enthusiasm is contagious. Already a strong classroom manager, she's seen student engagement increase at an amazing rate.

"They are quiet and working – it's kind of spooky."

Some of her fourth graders are reading on the first grade level, and she has become strategic about how she uses the software to help.

"I can share a passage through Learning Notes that we both can see on our notebooks. I can annotate or highlight and define a word they are struggling with. They read for a minute then we talk about it. Then they reread the text but they record themselves and hear themselves reading. That reflection helps to improve their fluency."

She also uses it to have the computer read back to them using the immersive reader function. It breaks down the syllables for them.

"I've seen the kids be more excited and not feel defeated," she said.

Learning in the Mall of America

Schools are, of course, the natural place for learning to occur, but one Minnesota mother will not wait for dyslexics to get the assistance and support they need. Recognizing dyslexia in the classroom requires state governments to act, and for state governments to act requires passionate advocacy. Rachel Berger, the mother of a young dyslexic student, founded Decoding Dyslexia, a national grassroots movement. To understand why she took action requires a little explanation.

Rachel has observed and been affected by dyslexia her entire life. Her father dropped out of school in the ninth grade and both of her brothers struggled with learning disabilities. Several years ago she noticed her own son was struggling and decided to have him diagnosed. While awaiting results, her father, then 68, also decided to be tested. He wanted to go back to school, but was worried that his IQ was so low that he might not succeed.

"He called a week later to tell me, 'I am not stupid. What I am is profoundly dyslexic," she recalled. "We were a day away from my son's diagnosis."

Her father noted that in the 1950s when he was in school they didn't know any better. Today, screening is easily accessible.

"I can't stand to watch my son and my grandson go through this without any help," her dad said.

Just as she thought, her son's test results came back positive for dyslexia.

"The schools were not set up to deal with him systemically. I could either watch him fail like I did my brothers, or I could step out of the school, and so I removed him and got him tutored. It came at a significant cost. I couldn't reconcile this pit of sickness when I would walk into the school. That motivated me to start Decoding Dyslexia Minnesota. We are going to have equitable education here in Minnesota."

Rachel began advising a Swedish company that uses cloud technology for an app that screens for dyslexia. One day, the company invited Mike Tholfsen to demo Microsoft's Learning Tools for its advisors, and Rachel still expresses shock when she recalls seeing the immersive reading technology.

"How on earth do we not know about this?" she

exclaimed.

Recognizing the awareness problem, she told Mike that every parent and student should know about it. Rachel worked with Mike to set up a pilot for a community night on dyslexia at the Microsoft Store. She invited students, parents, teachers and tutors she knew from the community to this event, and partnered with store employees to do interactive trainings on Learning Tools.

Suddenly, the Microsoft retail store at the gigantic Mall of America in Bloomington, Minnesota was promoting demos of Learning Tools for anyone and everyone. Ninety percent of the people that night had never been to a Microsoft Store in the past, and some families drove four hours just to get there. Now, this type of event is being planned broadly across all Microsoft Stores around the country.

"I know how to activate the WOMMA – the Word of Mouth Mom Army," she said.

Rachel also knows how to change state policy. There are now chapters of Decoding Dyslexia in every state, and they are ensuring that dyslexia is understood in state statutes and supported with resources and expertise. She writes:

> *We need to change our knowledge deficit on dyslexia and its implications both educationally and*

within the workplace. Too often individuals with dyslexia are labeled as "lazy, dumb, not working to potential" by teachers, employers or peers. When dyslexia goes unidentified in the educational setting (as it often does) beyond the appropriate window of remediation, the only option students have is to attend special education classes, which don't offer specific interventions to help the student close the literacy gap between them and their peers, much less ever catch up.

We need to be investing in tools for early identification and appropriate remediation. We can't afford to sideline individuals with dyslexia, their potential is too great and the impacts to society are too significant to continue to bear.

- - -

A 2017 study conducted by RTI International, an independent nonprofit research institute, found that Learning Tools students showed an average gain of 123.6 points in their reading comprehension scores, compared to the historical group's 89.2 points, and an average gain of 10 percentile points.

In the United Kingdom, the government's Department of Work and Pensions (DWP), which assists jobseekers, is experimenting with the read-aloud tools originated in the Learning Tools hack and now part of Microsoft's Edge browser. Many applicants lose time and money seeking help because they cannot access the online services due to their literacy level, vision impairment, English as a foreign language or a physical disability. Microsoft was able to create a number of personas of service users to illustrate the problems faced and built a demo which shows the power of accessibility tools.

In the demo, a person is able to find the contact details of the DWP Job Search program with read aloud. They are then connected to a Job Centre of the Future bot to use voice only to set up an appointment.

As this book was going to press, the team's executive sponsor, Eran Megiddo, emailed the entire hack team.

"The obsession and drive to understanding the user needs and the dedication and passion to bringing the capabilities to market, driving integrations, and awareness are outstanding. As I have been telling the team – let's keep pouring fuel on this fire: what can we do to further improve people's ability to read? In what other ways can technology help them overcome any challenges they may have? And what can we do to further broaden the reach and awareness of what we already have and what we are yet to deliver."

Conclusion

Years after the first Ability Hack, Jeff Ramos, who leads The Garage, paused to reflect on what's transpired. The Garage is a resource for Microsoft employees that supports and encourages problem solving in new and innovative ways and led the all up company hackathon since it began in 2014. The Ability Hack projects helped employees contribute beyond product strategy. Ability Hacks helped Microsoft people find meaning in their work and a sense of purpose – the hacks were personal.

"With a typical technology hackathon, project teams are recognized for the power of their idea and its potential impact," Jeff said. "In essence, it's an endorsement of their thinking. Some of these projects help us be more productive at work, as an example. With Ability Hack projects, the benefactor seems less anonymous and people can visualize the meaning and impact that a project has for another human being. Ability projects fundamentally help people live better lives by delivering more empowerment – that's what evokes the emotion and passion."

There are many lessons from these hacks, and we hope you will share yours, too #abilityhacks.

Among those we heard most frequently are:

- Great hacks must appeal to the heart and the mind – working on technology that is pushing the boundaries of what's possible, to the benefit of people with disabilities.

- Hackers love great technical problems. They want big nuts to crack– major challenges they must stretch to get their arms around.
- Hacks must have engaged customers, the real users.
- Successful hacks always highlight the power of a true team, and a strong leader(s) at the helm – developers, program managers, storytellers.
- A great hack almost certainly will attract serial hackers that come back year over year, and likely make first-time hackers serial hackers.

AFTERWORD

By Jenny Lay-Flurrie

Jenny Lay-Flurrie was named Chief Accessibility Officer in early 2016. She is the first deaf person to occupy a C-suite office. Ever.

Ten years ago, I came off stage after giving a presentation at a conference, and a member of the audience came up to me, vigorously shook my hand, excited and barely able to speak. The only words she could say were "you're so inspirational," which she said over and over and over. I smiled, nodded, gave her a ton of thanks and sat back down. I told my Dad about it back home in the UK, and he couldn't control his laughter. "What on earth are they finding inspirational about you, what on earth did you do on stage!?" I played it back in my mind. I had woven the word avocado into the speech (never compete with me on random word bingo), worn screamingly high heels, and my slides were the bomb. But none of that added up to inspirational for me. I couldn't figure it out!

The next conference, the same thing happened but this time I asked, "Why inspirational?"

"Because you do what you do, and you're deaf, that's so inspirational. You're so brave!"

I've had deafness since I was a teeny tot thanks to a combination of measles and genetics. I wasn't the only one in my family. In fact, in my house, if you had hearing you were the odd one out. You'd also walk in to a cacophony of sound. The stereo, record player and every radio in the house would be playing Salt 'N Pepa, Duran Duran or Adam Ant at full volume. My sister and I loved music but would never hear it well enough to remember to turn off the stereos. It was normal to be deaf. It was normal to have a disability.

However, disability wasn't normal outside of our wee home and isn't "normal" in much of society. Unemployment in the U.S. among people with disabilities remains twice that of the overall population, and the labor participation rate is only 17.9 percent for people with disabilities, compared with 65.3 percent for general population. And over 70 percent of disability is invisible. Add it all up and you're lucky if you see a person with a disability in the workplace. Something we're working hard to change and it's critical we do, as there are over one billion in the world with disabilities. It's the largest minority on the planet.

There are two big ways Microsoft can impact this. First, as an employer. Our early employee disability communities started in the 1990s, kicked off by passionate employees with blindness, deafness, ADD and parents to kids with autism. Disability has always been part of the fabric of Microsoft and in recent years we've focused hard to grow the talent pool. The simple truth is we need people with disabilities in the fabric of our company. Satya said it best when commenting on the third anniversary of our Autism Hiring Program: "Inclusive teams that value diverse perspectives and inclusive design principles will have the deepest impact in building products designed for everyone." By hiring talent with disabilities, regardless what role or position, we get valuable insight and experience that makes our whole approach richer, open ups our products to everyone, and builds a culture that lives and breathes disability inclusion.

The second way is through our technology and ensuring our products empower people with disabilities by prioritizing accessibility. Bill Gates kicked off our first accessibility team back in 1997 when he announced formation of Trustworthy Computing. The simplest definition of accessibility is how we ensure that products work for individual needs and preferences. In many ways accessibility is an engineering discipline on par with security and privacy and as important. The optimal way to achieve accessibility is through inclusive design, thinking through all the use cases as you take the magical ideas and turn them into nuts and bolts...or code. If you breathe it into your design process, you'll end up with a product that not only empowers cool people with disabilities, you may open doors to innovation – talking books, door handles, fluorescent light – the list is long. Or you could try and render a product accessible at the end of the cycle and face untold misery (and cost). Just imagine for a minute you're getting ready to open a freshly built condo building, and then hours before you cut the red ribbon, you realize the building is inaccessible. Cardboard ramps aren't great with 350 pound wheelchairs, and neither are Band-Aids on code.

Back in 2014, we were reenergizing our efforts on accessibility to live into the new mission of the company – empowering every person and organization on the planet to achieve more. I was chair of the disability employee group and working grass roots to harness the talents and expertise of our employees to help build better products. Then came the hackathon. I remember getting the email from Satya about Steve Gleason and getting stupidly excited. After the Super Bowl ad, I had been trying to find the right person to get an audience with Steve. He was an expert user of eye gaze, something we knew little about at the time, and an expert in ALS or MND, again something we knew little about. It was also personal. Two colleagues of mine had recently shared that they had been diagnosed with ALS and asked me to help.

I had the Ability Hack already underway with nine projects lined up. I added the eye gaze hack project and within three days of posting the project for Steve my inbox was full and people were literally chasing me down the corridor. None had experience in accessibility, in fact in the end, only two out of the 30+ folks on the team knew anything about accessibility. The rest came from diverse disciplines – from developer to engineer,

PM to marketer to photographer. Experts in Surface hardware, robotics, internal IT systems, online advertising, based in Redmond, China and more, male, female, disability, no disability ...you name it, we had it. And every single person was buzzed.

That week is still one of the best of my working life. Steve and his family drew you in within seconds. His first question to me in a conference room in Seattle was "how a deaf girl gets a music degree" (he'd done his homework), and most importantly, what football team did I support (his ManU, mine Aston Villa). He then patiently taught me how he uses his technology, over a series of hours and days. Three short days later, Steve had met with Satya, VPs from around the company, had his entire system overhauled by Surface engineering and support staff, and inspired a 30+ person strong hack team to work on technology that in the months and years to come would change his life. Thanks to the dedicated work of the Microsoft Research team, the many amazing people with ALS (PALS) who have helped us here in Seattle and, in more recent history, the Windows teams, that vision will (fingers crossed) become a reality for many thousands more living with ALS.

Learning Tools came around the following year. They were the first team in the tent on opening morning, determined to grab a good spot, and throughout the event you could see sparks fly as the team tackled every problem one by one. I remember sitting down with Jeff Petty for a quick run through and realizing the absolute potential of their approach. This could be life changing for kids, heck all of us. Three years later, the product has just hit 13M unique monthly users and growing and is changing the landscape for people, kids with dyslexia.

In 2014, Microsoft had 10 Ability Hack projects and 75 people hacking for disability. Last year, we had 150 Ability Hack projects and over 850 people. I have no idea what summer 2018 will bring, but I do know that Ability Hack has become a crucial part of our strategy at Microsoft.

In 2016, I became the Chief Accessibility Officer and started the journey of driving accessibility systematically into the culture and fabric of Microsoft along with an army of awesome humans across the company. To make it part of our DNA and motivate 100K+ employees to live and breathe inclusive and accessible design and empower both our customers and ourselves. My learning through the last four years of Ability Hack is there is no better way to motivate 100+K rather nerdy souls than dedicating two entire days to tackling tough real human problems.

Every year, teams submit their ideas and project names ahead of the hackathon. By the end, it's clear which have elements of genius and which are early on their journey, but all walk away with more than they came in with. At a minimum, every person gains a higher understanding of disability, of etiquette, language, accessibility principles and more. Last year we even put up signs on whiteboard and a sign on every pillar and whiteboard that said "An Inaccessible Hack is Whack," with details on how to ensure every hack was inclusive. All of that goes back with them when they leave the tent. It enters hearts and minds. It builds DNA.

Accessibility is a journey. Whether you're planning to build your own Ability Hack, or wondering where to start in your own journey, my recommendation to you all is to take a few steps first to learn and explore how to create an inclusive and accessible culture. Some quick hints and tips:

1. Know how accessible and inclusive you are. Ask your manager, peers, employees. Test your products. Take a look at your events. Is your hack space accessible, your email inclusive across the spectrum of disability? Remember that 70% of disability is invisible and you may not know that you have a person on your team with a disability. If you're reading this and you don't know how accessible your app, website, or product is – it's not accessible. End of story.

2. Get educated. Read any of the books, websites and standards, or Ted Talk yourself into academic excellence.

3. Build a plan to ensure accessibility is embedded

from design forward. Figure out your priorities. Reach out to the oodles of experts, NGOs and communities focused on accessibility and hiring talent with disabilities. Allow them to guide you. Set a goal. Overachieve, I dare you.

4. Motivate your culture to think about all their users. Accessibility is cool! It's innovation, inclusion and so much more. Make sure it's connected to your strategic goals and not a side line project. Inclusion is too important to be benched.

5. Be people and "empathy led" in your approach, listen and learn from the experts – people with disabilities — and follow the principles of inclusive design. Get connected to your disability employee group and/or hire a person with a disability. Worried about the mechanics of hiring talent with a disability? Do it anyway. You'll learn quickly.

6. Be careful of the "soft bigotry of low expectations." No one should be inspired by someone with a disability getting up in the morning, going to school or having a job. Be inspired by our latest hire through the autism program, who coded so brilliantly and quickly during assessment, the hiring managers asked for a two hour break so they could figure out what he had done.

Back to that stage. Ten years on, I still come off stage and meet overexcited, wonderful humans who tell me I'm inspiring and sometimes brave. Inspiration is important, and every time I'm humbled by the conversation and impact a few words have had. I really do believe that one day disability will be the norm. But let's be clear, I'm not inspirational because of my disability. I'm not brave (just put a spider next to me and see what happens, or watch my hands shake before I go on a stage). I'm inspirational because I represent a company with an incredible mission to empower every person and organization on the planet to achieve more. Because I work every day with a community both internal and external to Microsoft that believes nothing is impossible and is prepared to put every nerdy ounce of energy into making a new reality. I'm inspiring because I'm a sassy, smart, incredibly funny, moderately humble human who is prepared to kick ass to make a difference in this world. And I happen to be deaf.

APPENDIX

Microsoft Accessibility Features

by Guy Barker

Guy has been a developer at Microsoft for over 20 years. He has spent the last 12 working on Microsoft Accessibility, in Windows and in our core accessibility team.

For many years as a developer at Microsoft, a couple of things have intrigued me. One of those is how can I leverage today's technologies to plug a gap in what's available to someone? And I'm far from alone in that exploration, as all the people involved in the many hackathons happening these days clearly demonstrate. For me, one key moment was when I met a student who wasn't able to physically speak and would write down what he wanted to communicate on a piece of paper. Given how easily a dev could leverage the ink recognition and text-to-speech features in Windows, it took only minutes for me to build a prototype ink-to-speech app for the student to try out. Another key moment for me was being with my mum as she tried using a foot switch to control a Windows device. The fascinating question of what I can do to help people interact with their devices has never left me.

Another thing that's intrigued me: What can I do to

help all devs build Windows apps that are intuitively and efficiently accessible to everyone? Whether an app provides a means of connecting with other people or a means of employment, everyone deserves to benefit from today's powerful technologies. While it's natural for devs to build apps that they as individuals can interact with, a true sense of pride comes with being creative enough to build an app that all people can efficiently use.

And so we must ask:

You've built something important. Something that matters. You might have built it during a hackathon, and you'll guide it along its path to becoming a part of a shipping Microsoft product, or it might be one of your product team's spec'd features. Either way, you know it will impact people's lives, and you want every person to be able to benefit from what you've built regardless of how they interact with their device. So when any person asks you if they can use your app, what will you say to them?

This appendix describes many accessibility features that customers use to leverage all the functionality in accessible software products. These accessibility features include Eye Control, Windows Speech Recognition, On-Screen Keyboard, Magnifier, the Narrator screen reader, Captioning, Color Filters and many more. When one of, or some combination of, these forms of input and output are being used, your customers are depending on you to provide an intuitive and efficient experience for accessing everything your app has to offer.

All app builders want to deliver a great experience for every person on the planet, so how does this affect the code that devs write?

Many UI frameworks do a ton of work on your behalf to make your features accessible to everyone. If you're using standard controls and widgets in UWP XAML, WPF, WinForms, Win32 or HTML hosted in Edge, then in many ways, the UI will be accessible by default. For example, a button will often be fully accessible to customers using only the keyboard, and to customers using the Narrator screen reader, and will respect appropriate system colors when a high contrast theme is active. All that, just because you used a button.

By default, always use standard controls that come with the UI framework, even when you're customizing the visual representation of the control. For example, if you have UI that looks like a towhee but behaves like a button, then style a standard button to look like a towhee. Don't instead render an image of a towhee in the UI, and then try to add support for the keyboard and for programmatic access. Take advantage of how the UI framework can help you build fully accessible UI in the shortest time possible.

Now, say that accessibility was considered during the

design phase of your UI, and you've used standard controls wherever practical in your app. As a result, you've got a solid head start on delivering a great experience for everyone. There will still be occasions when your customers will be depending on you to take specific additional steps to enable them to leverage your great features.

For example, you have your button that visually shows only an image of a towhee. If there's no equivalent text string programmatically associated with the button, then a screen reader such as Narrator won't be able to inform your customers as to which button they've encountered. So in this situation, as part of building the UI, you'd spend a moment adding a localized string that will get announced by a screen reader when your customer encounters the button, even though the string isn't displayed visually in the app. This sort of step can be as quick to do as it would be to add a visual string, and it can make the difference between the app being usable and not usable.

And before you ask, no, the frameworks aren't at a point where AI is used to recognize that the button's showing a towhee, and not (say) a grosbeak. You know the exact meaning of everything in your UI, and you can robustly convey that meaning in the manner most helpful to all your customers.

The point here is that often there are quick 'n' easy steps you can take to enhance the accessibility of your

UI, and those steps have a huge impact on your customers. To learn more about how you can help your customers, visit http://aka.ms/devenable.

I'm really pleased to have had the opportunity to share some additional knowledge on building accessible UI at http://aka.ms/uiablog, so you can find additional tips there. And given that no team can deliver a truly accessible app without being aware of the app's programmatic representation, another excellent resource is http://aka.ms/uiaintro. (I'd probably say that resource was excellent even if I wasn't in it.)

The UIA resource just mentioned demos a couple of Windows SDK tools that can help you build UI that's accessible to assistive technologies like screen readers and magnifiers. Another incredibly important tool you can use to help verify the accessibility of your app is called the "keyboard." Some of you might be familiar with this tool, as it's what you may spend all your days writing code with. A huge number of customers use only the keyboard for input, and so rely on you delivering a great experience when that's the only input device being used. So update your development processes to regularly try accessing all your functionality using only the keyboard. If you feel you don't have time for that, then you already know your customers won't be getting the experience they deserve. That's because using the keyboard to leverage all that your app has to offer should be quick.

Design for everyone, use standard controls by default, and verify the results. Where required, take those specific steps provided by the UI framework to round out the accessible experience that your customers depend on. And by doing so, feel the same excitement that all app builders do when they read about the accessibility features described in this appendix and realize how their app can be used by customers through Eye Control, Windows Speech Recognition, On-Screen Keyboard, Magnifier, Narrator...

As you learn about the features in this appendix, consider how each feature might help someone interact with your app. And as you build up your processes to design, build and test products that every person can efficiently use, always keep in mind the rationale for doing so. It's a question of fundamental fairness. No person should be prevented from being connected with others, or from being employed, simply because apps block them from those core aspects of life. And what's more, with the tools at your disposal, it's practical for you as an individual who cares about quality for all to build an accessible app.

The goal is to reach the point where any person can ask you whether they can use your app, and with confidence, you can reply: "Yes. You can."

Key Microsoft Accessibility Features

We keep adding to the list of and improving our existing accessibility features. What is listed below is a snapshot, please check out www.microsoft.com/accessibility to see the most current product features. (Download the digital version of the book for additional links.)

Vision

Accessibility Checker

Accessibility Checker is now easily discoverable in Word, Excel, PowerPoint, OneNote, Outlook and Visio. The Accessibility Checker analyzes your material and provides recommendations alongside your document, helping you understand how to fix errors and create more accessible content over time.

We enhanced the Accessibility Checker to streamline the process of creating quality content that is accessible to people with disabilities. Now, the Accessibility Checker identifies an expanded range of issues within a document, including low-contrast text that is difficult to read because the font color is too similar to the background color. The checker also includes a recommended action menu and utilizes AI to make intelligent suggestions for improvements—like suggesting a description for an image—making it easier to fix flagged issues from within your workflow.

Select 'Check Accessibility' under the 'Review' tab to get

started.

Use Case: Before sharing content, you can run the accessibility checker to find and fix any issues that might make your content difficult for people with disabilities to use. For people with visual disabilities Accessibility Checker will identify heading structure, alt text, order of content, etc.

Alternative Text Tool

Alternative Text (Alt Text) provides a description for a photo or object so that a person who is blind can understand the context. The feature to manually add or change Alt Text is now conveniently located on the right click menu in Office.

Use Case: The new placement of the alt text feature makes it easy to check auto generated descriptions for accuracy and to provide alt text on charts, graphs and other media.

Automatic Alternative Text

Automatic Alterative Text (Alt Text) leverages intelligent image analysis powered by the Microsoft Computer Vision Cognitive Service to automatically give suggestions for alt-text descriptions for images in both PowerPoint and Word. Auto Alt Text helps save you time and ensures your media-rich PowerPoint presentations, Word documents and Outlook emails can be consumed and understood by people with blindness or visual impairments.

Use case: For people who are blind or have low vision, automatic Alt Text provides descriptions of images without a content creator needing to add it. The Alt Text is tagged with a message informing the user that it has been automatically generated and provides a confidence rating regarding the accuracy of the image description. Through leveraging machine learning algorithms, the alt text will continue to improve over time.

Color Filters

If it's hard to see what's on the screen, apply a color filter. Color filters change the color palette on the screen and can help you distinguish between things that differ only by color.

Use case: With color filters applied, differences between colors are easier to detect for people with color blindness.

Consistent Keyboard Shortcuts

Keyboard shortcuts (e.g., Ctrl + C to Copy) are the same across the whole OS regardless of version.

Use case: With universal shortcuts, people with a vision disability only have to memorize a single set of tools, which reduces the amount of mental mapping required to operate Windows and the number of keystrokes necessary to complete a task.

Edge Reading Mode

Sometimes you just want to cut out the distractions and get immersed in an article or story. Reading mode is now in Edge, which removes ads, banners and navigation elements from the page and changes the background to an easier to view color.

Guided Setup

There is now Cortana-guided audio-based initial setup of a Windows 10 device with no touch or mouse required.

Use case: This out of the box feature during initial device configuration allows you to perform setup using voice, Narrator or Keyboard only.

High Contrast

High Contrast increases the color contrast of text and images on your screen making them easier to identify.

- In **Windows**, each High Contrast theme can be customized to suit your needs and tastes.
- In **Office**, High Contrast in the ribbon has increased contrast, has fewer icons with grey gradients, and icons have clearer outlines that

make it easier for people to identify them at a glance.

- In **Xbox**, you can enable High Contrast from the Ease of Access settings on your Xbox One console.

Use case: People with certain vision impairments, such as cataracts, rely on High Contrast themes to see apps and content with less eye strain. With the High Contrast mode turned on, the icons on the Office ribbon are more visible to someone with reduced contrast sensitivity. Try it out! Press Left Alt + Left Shift + Print Screen on your keyboard.

Keyboard Only

We have enabled the ability to navigate through the Operating System or Office Suite using only the keyboard (no mouse, no touch).

Use case: Certain users navigate a computer using the keyboard rather than the mouse, as an example, while using a screen reader, working with a Braille keyboard or other Assistive Technology.

Live Tile Sizes

In Windows and Apps, tiles can be changed to multiple sizes for prioritization and easier viewing. Live tiles in Windows 10 display information that is useful at a glance without opening an app.

Use case: The News tile displays headlines while the Weather tile displays the forecast. You can rearrange, resize and move these tiles to make them work better for you.

Magnifier

Magnifier is a tool that enlarges part—or all—of your screen so you can see words and images better. It comes with a few different settings, so use it the way that suits you best.

Mono Audio

By default, most stereo audio experiences send some sounds to the right channel or speaker and some sounds to the left channel. Windows supports mono audio, so that you can send all sounds to both channels.

Use case: This allows users to hear all sounds from the PC in one ear while listening to people around them with the other.

Narrator

Narrator is Microsoft's built-in screen reader for Windows and Microsoft applications (e.g. Office365). It enables you to interact with the computer without viewing a screen and provides command and control of your device using a keyboard, controller or with a range of gestures on touch supported devices. Narrator comes with support for 27 languages, enables you to install more, and supports multilingual reading.

Office Lens

Office Lens is a portable scanner with built-in image cropping, optical character recognition (OCR) and contact syncing. Files can be shared via the cloud to other devices and software (e.g. OneNote, Word). With Immersive Reader you can now have printed documents, whiteboards, handwriting, business cards and more read out loud.

Use case: Use it to take pictures of receipts, business cards, menus, whiteboards or sticky notes—then let Office Lens crop, enhance and save to OneNote.

Office Templates

The most popular templates in Office 365 have accessibility built in. They help you make your content accessible to everyone in your organization.
Want to make your own templates? Several of these templates are not just for you to use, but also guide you through making your own accessible templates. Open the templates to discover tips and how-to instructions for making templates for your organization to use.

Read Aloud

Read Aloud in the Edge browser can now read websites, PDFs and ePUBs aloud. All you need to do is press the "Read aloud" button located at the top-right corner of the browser, sit back, relax and listen. Edge will also highlight the words being read. You can also right click and choose "Read Aloud".

Use case: Reading comprehension of long documents is made easier when words are highlighted while read out loud. This feature is a great inclusive design tool for reading long, complicated text, when multi-tasking, or when folks get tired at the end of the day.

Seeing AI

Seeing AI is a free iOS app that narrates the world around you. Designed for the blind community, this research project harnesses the power of AI to describe people, text and objects.

- **Short Text:** Speaks text as soon as it appears in front of the camera
- **Documents:** Provides audio guidance to capture a printed page and recognizes the text, along with its original formatting
- **Products:** Gives audio beeps to help locate barcodes and then scans them to identify products
- **Person:** Recognizes friends and describes

people around you, including their emotions
- **Scene:** An experimental feature to describe the scene around you
- **Currency:** Identify currency bills when paying with cash
- **Light:** Generate an audible tone corresponding to the brightness in your surroundings
- **Color:** Describes the perceived color
- **Handwriting:** Reads handwritten text

Tell Me

The 'tell me what you want to do' feature in the Office ribbon makes it easier for you to discover the difficult-to-find capabilities in Office by simply typing into the search box. 'Tell me' will take you directly to the feature/function you want to use without requiring you to search the ribbon or menu.

Use case: 'Tell me' helps people find features or functions without having to look through multiple ribbons containing numerous features. This can be helpful to people who are blind or have low vision, have cognitive disabilities, aging memory loss, or to prompt anyone that can't remember where a specific feature is located.

Text Scaling

For Windows and Apps, Text Scaling provides easy scaling of text, apps and other items within existing resolution.

Use case: If text and other items on the desktop are too small, you can make them larger without changing the screen resolution or turning on Magnifier.

Themes in Office 365

Themes in Office 365 allows users to pre-set color and text combinations within the Office Suite without turning on High Contrast. If you feel like the Office color scheme is too bright or you need more contrast, you can change the Office theme for all your Office programs from your account settings.

Use case: Changing color schemes can make it easier to see Office 365 applications, potentially reducing eye strain and increasing focus.

Touch Feedback

Touch Feedback shows visual feedback (ripple) when contact is made with a touchscreen and can be darkened for greater visibility.

Use case: Provides greater focus of where touch contact is made with a screen, useful for more easily tracking contact, especially useful for a person who may be learning how to interact with a touchscreen for the first time.

Windows Hello

Windows Hello is a more personal way to sign in to your Windows 10 devices by using facial recognition or your fingerprint. You'll get enterprise-grade security without having to type in a password.

Use case: Windows Hello provides a secure way to log into a computer for people who easily forget passwords or require assistance entering a password.

Cognitive

Accessibility Checker

Accessibility Checker is now easily discoverable in Word, Excel, PowerPoint, OneNote, Outlook and Visio. The Accessibility Checker analyzes your material and provides recommendations alongside your document, helping you understand how to fix errors and create more accessible content over time.

We enhanced the Accessibility Checker to streamline the process of creating quality content that is accessible to people with disabilities. Now, the Accessibility Checker identifies an expanded range of issues within a document, including low-contrast text that is difficult to read because the font color is too similar to the background color. The checker also includes a recommended action menu and utilizes AI to make intelligent suggestions for improvements—like suggesting a description for an image—making it easier to fix flagged issues from within your workflow.

Use case: Before sharing content, you can run the accessibility checker to find and fix any issues that might make your content difficult for people with disabilities to use. For people with visual disabilities Accessibility Checker will identify heading structure, alt text, order of content, etc.

Captioning

Windows 10 closed captions will display text of the words spoken in the audio portion of a video, TV show or movie. To customize closed captions, go to the Ease of Access Center under Settings.

Office365 supports closed captions and subtitles that are embedded in video files. Authors can now associate closed caption files with their audio recordings or audio files added from their local drive or OneDrive/OneDrive for Business. Office365 authors can also associate closed caption files with any video files uploaded from their local drive or from OneDrive for Business.

Xbox supports closed captioning. It will display captioning that developers put into games during development, and there is captioning capability in Xbox settings.

Use Case: Adding closed captions makes your

presentation accessible to a larger audience, including people with cognitive disabilities and learning differences. Closed captions are an example of Universal Designed Learning (UDL) and provide several ways for students to access content. Closed captions increase learning comprehension in all students in a classroom.

Consistent Keyboard Shortcuts

Keyboard shortcuts (e.g., Ctrl + C to Copy) are the same across the whole OS regardless of version.

Use case: With universal shortcuts, people only have to memorize a single set of tools, reducing the amount of mental mapping required to operate Windows and the number of keystrokes necessary to complete a task.

Dictation

Dictate for Word, PowerPoint, Outlook 2016 Desktop for Office 365 and OneNote for Windows 10 converts speech to text using the state-of-the-art speech recognition behind Cortana and Microsoft Translator. It includes:

- Highly accurate speech to text
- Two modes of punctuations: Auto and manual (with commands like "Question mark," Period," Exclamation mark" and "Comma"
- Visual feedback to indicate speech is being processed

- Just two keystrokes to enable the dictation feature: Windows Key + H

Edge Reading Mode

Sometimes you just want to cut out the distractions and get immersed in an article or story. Reading mode is now in Edge, which removes ads, banners and navigation elements from the page and changes the background to an easier to view color. Edge Reading Mode now has immersive reading features including Read Aloud, theme colors, spacing mode, syllables, parts of speech highlighting.

Editor

Editor provides advanced proofing and editing service. Leveraging machine learning and natural language processing—mixed with input from our own team of linguists—Editor makes suggestions to help you improve your writing.

How? Editor provides visual proofing cues to distinguish at a glance between edits for spelling (red squiggle), grammar (blue double underline) or writing style like wordiness, redundancy or etiquette (gold dotted line).

- For advanced spelling support, a list of suggested words with synonyms is provided as well as the ability to have the word read out loud.

- For writing style, Editor helps you simplify and streamline written communications by flagging unclear phrases or complex words, such as recommending "most" in place of "the majority of." It will reinforce first person disability language by flagging phrases like "blind person" with "a person who is blind."

Focus Assist

Focus assist allows you to turn off notifications in Windows anytime you need to get in the zone. You will not receive notifications from any of your installed apps.

Use case: For people with cognitive disabilities like ADHD, TBI, autism, and people with mental health disabilities such as anxiety may benefit from reducing distraction of notifications. Eliminating moving objects such as notifications may also reduce stress and help people focus.

Game Chat Transcription in Xbox

Xbox has speech-to-text and text-to-speech translation through game chat.

Use case: Clear communication is important to the success of multiplayer games. As more people are added to a game chat, however, it can become virtually impossible to differentiate voices and conversations.

Translating those spoken words into written text, in real time, makes it easier to keep track of all the dialogue as it happens, increasing comprehension and clearing communication channels.

Note Game transcription is available on Xbox One and Windows 10 PCs in games that support it. See the following link in 'for more information' for a list of games that support it.

Haptic Feedback in Xbox

Xbox has touch-based feedback given as vibration through a controller.

Use case: Allows for successes or errors to be reinforced in a tactile way in addition to visual or audio cues (e.g. vibration as a result, as damage or as proximity to success).

Inking

Windows Ink is a set of pen-driven experiences that help you set your ideas in motion with your pen. You can draw freehand to make annotations, highlight text or quickly draw shapes.

With a touch-enabled device, you can draw with a mouse, your finger or a digital pen. The exact features available depend on the type of device you're using and whether you're an Office 365 subscriber. In some apps, for instance, you can replay ink or convert ink to shapes or math.

Use case: Many people with a cognitive disability or dyslexia can get more information down using the pen than a keyboard. The Microsoft pen allows inking and erasing in an intuitive and simple way.

Microsoft Learning Tools

Microsoft Learning Tools are free tools that implement proven techniques to improve reading and writing for people regardless of their age or ability.

- **Improves comprehension:** Tools that read text out loud, break words into syllables, increase spacing between lines and letters, highlighting parts of speech (nouns, verbs, adjectives), support line focus and picture dictionary.
- **Encourages independent reading:** A teaching aid that helps teachers support students with different abilities.

Learning Tools capabilities, including the Immersive Reader, are built in across Office and the Edge browser and supported in many languages.

- OneNote - Desktop, Windows 10 app, Web, iPad, Mac,
- Word – Desktop, Online, iPad, Mac
- Outlook – Desktop, Web

- Office Lens iOS
- Edge – ePub, PDF, Web, Reading View

Mono Audio

By default, most stereo audio experiences send some sounds to the right channel or speaker and some sounds to the left channel. Windows supports mono audio, so that you can send all sounds to both channels, so you don't miss anything if you have partial hearing loss or deafness in one ear.

Nightlight

Nightlight allows for screen adjustments with blue light filtering to reduce stimuli, especially at night.

Use Case: Your display emits blue light—the kind of light you see during the day—which can keep you up at night. To help you get to sleep, turn on Nightlight and your display will show warmer colors at night that are easier on your eyes.

Office Templates

The most popular templates in Office 365 have accessibility built in. They help you make your content accessible to everyone in your organization.
Want to make your own templates? Several of these templates are not just for you to use, but also guide you through making your own accessible templates. Open

the templates to discover tips and how to instructions for making templates for your organization to use.

Office Lens

Office Lens is a portable scanner with built-in image cropping, optical character recognition (OCR) and contact syncing. Files can be shared via the cloud to other devices and software (e.g. OneNote, Word). With Immersive Reader you can now have printed documents, whiteboards, handwriting, business cards and more read out loud and space out the text.

Use Case: Use it to take pictures of receipts, business cards, menus, whiteboards or sticky notes—then let Office Lens crop, enhance and save to OneNote.

PowerPoint Designer

PowerPoint Designer improves your slides by automatically generating design ideas that you can choose from. While you're putting content on a slide, Designer works in the background to match that content to professionally designed layouts.

Read Aloud

Read Aloud in the Edge browser can now read websites, PDFs and ePUBs aloud. All you need to do is press the "Read aloud" button located at the top-right corner of the browser, sit back, relax and listen. Edge will also highlight the words being read.

Use Case: Reading comprehension of long documents is made easier for people with learning disabilities such as dyslexia when words are highlighted while read out loud. This feature is a great inclusive design tool for reading long, complicated text, when multi-tasking, or when folks get tired at the end of the day.

Remappable Buttons

Mappable buttons allow a game player with a physical disability to have custom buttons for their preferred input to Xbox games. This allows someone who cannot press all the buttons to participate. This is especially powerful when used in conjunction with Co-pilot mode.

Researcher

Researcher helps you find and incorporate reliable sources and content for your paper in fewer steps. You can search the Internet right within your Word document and you can explore material related to your topic and add it—and its properly formatted citation— in one action.

Use Case: Eliminate distractions and stay on task by conducting web searches while writing a document in Word. With automatic citation, you can teach students how to be good digital citizens and not plagiarize others' work.

Simplify Windows

In the Ease of Access settings, you can disable window animations in Windows 10, transparency and hidden scroll –bars, throughout the length and breadth of the OS.

Use Case: For people with cognitive disabilities like ADHD, TBI, autism, and people with mental health disabilities such as anxiety may benefit from eliminating animations to reduce distractions, negative emotional/behavioral responses and help them stay on task.

Smart Lookup

The Insights pane, powered by Bing, offers more than just definitions. By selecting a word or phrase and launching Smart Lookup, Bing can show you more information, definitions, history and other resources related to that word or phrase.

Select a word or phrase, right-click it and choose Smart Lookup. The insights pane will open with definitions, Wiki articles and top related searches from the web.

Use Case: Smart Lookup allows a person to rapidly access further information from within the Office suite.

Sway (AI)

Create and share accessible reports, presentations, personal stories and more. Sway has accessibility built in to make sure everyone can access your content.

Start by adding your own text and pictures, search for and import relevant content from other sources, and then watch Sway do the rest. With Sway, there's no need to spend lots of time on formatting. Its built-in design engine takes care of making your creation look its best.

Tell Me

The 'tell me what you want to do' feature in the Office ribbon makes it easier for you to discover the difficult-to-find capabilities in Office by simply typing into the search box. 'Tell me' will take you directly to the feature/function you want to use without requiring you to search the ribbon or menu.

Use Case: 'Tell me' helps people find features or functions without having to look through multiple ribbons containing numerous features. This can be helpful to people who are blind or have low vision, have cognitive disabilities, aging memory loss, or to prompt anyone that can't remember where a specific feature is located.

Text Suggestions in Windows 10

Text suggestions in Windows 10 allows users to get suggested words while they are typing. This is also known as "word prediction." Txt suggestions work with the software keyboard and hardware keyboard and can work with any application in Windows 10.

Themes in Office 365

Themes in Office 365 allows users to pre-set color and text combinations within the Office Suite itself without turning on High Contrast. If you feel like the Office color scheme is too bright or you need more contrast, you can change the Office theme for all your Office programs from your account settings

Use Case: Changing color schemes can make it easier to see Office 365 applications, potentially reducing eye strain and increasing focus.

Touch Feedback

Touch Feedback shows visual feedback (ripple) when contact is made with a touchscreen and can be darkened for greater visibility.

Use Case: Provides greater focus of where touch contact is made with a screen, useful for more easily tracking contact especially for a person who may be learning how to interact with a touchscreen for the first time.

Translator

As you speak, the add-in powered by the Microsoft Translator live feature allows you to display subtitles directly on your PowerPoint presentation in any one of more than 60 supported text languages. This feature can also be used for audiences who are deaf or hard of hearing.

- **Live subtitling:** Speak in any of the 10 supported speech languages – Arabic, Chinese (Mandarin), English, French, German, Italian, Japanese, Portuguese, Russian and Spanish – and subtitle into any one of the 60+ text translation languages.
- **Customized speech recognition:** Optionally customize the speech recognition engine using the vocabulary within your slides and slide notes to adapt to jargon, technical terms, product or place names, etc. Customization is currently available for English and Chinese.
- **Personal translations:** Share a Quick Response (QR) code or five letter conversation code and your audience can follow along with your presentation, on their own device, in their chosen language.
- **Multi-language Q&A:** Unmute the audience to allow questions from the audience on their device in any of the supported languages (10 for spoken questions, 60+ for written ones)
- **Inclusivity through Accessibility:** Help

audience members who are deaf or hard of hearing follow the presentation and participate in the discussion.

- **Translate your presentation while preserving the slide formatting:** Next to the "Start Subtitles" icon, the "Translate Slides" button allows presenters to translate their whole presentation while preserving its formatting.

Windows Hello

Windows Hello is a more personal way to sign in to your Windows 10 devices by using facial recognition or your fingerprint. You'll get enterprise-grade security without having to type in a password.

Use Case: Windows Hello provides a secure way to log into a computer for people who easily forget passwords or require assistance entering a password.

Hearing

Accessibility Checker

Accessibility Checker is now easily discoverable in Word, Excel, PowerPoint, OneNote, Outlook and Visio. The Accessibility Checker analyzes your material and provides recommendations alongside your document, helping you understand how to fix errors and create more accessible content over time.

We enhanced the Accessibility Checker to streamline the process of creating quality content that is accessible to people with disabilities. Now, the Accessibility Checker identifies an expanded range of issues within a document, including low-contrast text that is difficult to read because the font color is too similar to the background color. The checker also includes a recommended action menu and utilizes AI to make intelligent suggestions for improvements—like suggesting a description for an image—making it easier to fix flagged issues from within your workflow.

Use Case: Before sharing content, you can run the accessibility checker to find and fix any issues that might make your content difficult for people with disabilities to use. For people with visual disabilities Accessibility Checker will identify heading structure, alt text, order of content, etc.

Captioning

Windows 10 closed captions will display text of the words spoken in the audio portion of a video, TV show or movie. To customize closed captions, go to the Ease of Access Center under Settings.

Office365 supports closed captions and subtitles that are embedded in video files. Authors can now associate closed caption files with their audio recordings or audio files added from their local drive or OneDrive/OneDrive for Business. Office 365 authors can also associate closed caption files with any video files uploaded from their local drive or from OneDrive for Business.

Xbox supports closed captioning. It will display captioning that developers put into games during development, and there is captioning capability in Xbox settings.

Use Case: Adding closed captions makes your presentation accessible to a larger audience, including people with cognitive disabilities and learning differences. Closed captions are an example of Universal Designed Learning (UDL) and provide several ways for students to access content. Closed captions increase learning comprehension in all students in a classroom.

Game Chat Transcription

Xbox has speech-to-text and text-to-speech translation through game chat.

Use Case: Clear communication is important to the success of multiplayer games. As more people are added to a game chat, however, it can become virtually impossible to differentiate voices and conversations. Translating those spoken words into written text, in real time, makes it easier to keep track of all the dialogue as it happens, increasing comprehension and clearing communication channels

Mono Audio

By default, most stereo audio experiences send some sounds to the right channel or speaker and some sounds to the left channel. Windows supports mono audio, so that you can send all sounds to both channels.

Use Case: This allows users who have partial hearing loss or deafness in one ear to hear all sounds from the PC so they don't miss anything.

Office Templates

The most popular templates in Office 365 have accessibility built in. They help you make your content accessible to everyone in your organization.
Want to make your own templates? Several of these templates are not just for you to use, but also guide you through making your own accessible templates. Open the templates to discover tips and how to instructions for making templates for your organization to use.

PowerPoint Designer

PowerPoint Designer improves your slides by automatically generating design ideas that you can choose from. While you're putting content on a slide, Designer works in the background to match that content to professionally designed layouts.

Tell Me

The 'tell me what you want to do' feature in the Office ribbon makes it easier for you to discover the difficult-to-find capabilities in Office by simply typing into the search box. 'Tell me' will take you directly to the feature/function you want to use without requiring you to search the ribbon or menu.

Use Case: 'Tell me' helps people find features or functions without having to look through multiple ribbons containing numerous features. This can be helpful to people who are blind or have low vision, have cognitive disabilities, aging memory loss, or to prompt anyone that can't remember where a specific feature is located.

Translator

As you speak, the add-in powered by the Microsoft Translator live feature allows you to display subtitles directly on your PowerPoint presentation in any one of more than 60 supported text languages. This feature can also be used for audiences who are deaf or hard of hearing.

- **Live subtitling:** Speak in any of the 10 supported speech languages – Arabic, Chinese (Mandarin), English, French, German, Italian, Japanese, Portuguese, Russian and Spanish – and subtitle into any one of the 60+ text translation languages.

- **Customized speech recognition**: Optionally customize the speech recognition engine using the vocabulary within your slides and slide notes to adapt to jargon, technical terms, product or place names, etc. Customization is currently available for English and Chinese.
- **Personal translations**: Share a Quick Response (QR) code or five letter conversation code and your audience can follow along with your presentation, on their own device, in their chosen language.
- **Multi-language Q&A**: Unmute the audience to allow questions from the audience on their device in any of the supported languages (10 for spoken questions, 60+ for written ones)
- **Inclusivity through Accessibility**: Help audience members who are deaf or hard of hearing follow the presentation and participate in the discussion.
- **Translate your presentation while preserving the slide formatting:** Next to the "Start Subtitles" icon, the "Translate Slides" button allows presenters to translate their whole presentation while preserving its formatting.

Visual Notifications

In Windows, notifications disappear five seconds after they appear. This might not be enough time for you to notice them if you have difficulty seeing or hearing. You can increase the time a notification will be displayed for up to five minutes.

Mobility

Accessibility Checker

Accessibility Checker is now easily discoverable in Word, Excel, PowerPoint, OneNote, Outlook and Visio. The Accessibility Checker analyzes your material and provides recommendations alongside your document, helping you understand how to fix errors and create more accessible content over time.

We enhanced the Accessibility Checker to streamline the process of creating quality content that is accessible to people with disabilities. Now, the Accessibility Checker identifies an expanded range of issues within a document, including low-contrast text that is difficult to read because the font color is too similar to the background color. The checker also includes a recommended action menu and utilizes AI to make intelligent suggestions for improvements—like suggesting a description for an image—making it easier to fix flagged issues from within your workflow.

Use Case: This allows users who have partial hearing loss or deafness in one ear to hear all sounds from the PC so they don't miss anything.

Consistent Keyboard Shortcuts

Keyboard shortcuts (i.e. Ctrl + C to Copy) are the same across the whole OS regardless of version.
Use Case: This allows users who have partial hearing loss or deafness in one ear to hear all sounds from the PC so they don't miss anything.

Copilot

Want to use more than one controller on the same profile? The new "copilot" mode on Xbox One will let you do just that. Copilot for Xbox allows two physical controllers to be treated as one digital controller input.

Use Case:

1. Use two controllers simultaneously - one person can use two controllers at the same time (e.g. one in each hand, one with a hand the other with a foot, one with hand the other with a chin, etc.)
2. Play collaboratively - two people can share a single character/vehicle with controls divided between players and changeable as needed, leading to true cooperative play.
3. Work interchangeably - one person is in primary control while the second steps in as needed to assist with specific tasks (e.g. teaching a child to play, assisting someone over an especially difficult area, etc.)

Cortana

Cortana is Microsoft's virtual digital assistant. It is available as an app for Windows phone and iPhone. Bring your personal assistant to your phone to help you keep track of the important stuff wherever you are, across your devices. Set a reminder on your PC to pick something up at the store and Cortana will alert you on your phone when you get there.

Dictation

Dictate for Word, PowerPoint, Outlook 2016 Desktop for Office 365 and OneNote for Windows 10 converts speech to text using the state-of-the-art speech recognition behind Cortana and Microsoft Translator. It includes:

- Highly accurate speech to text
- Two modes of punctuations: Auto and manual (with commands like "Question mark," "Period," "Exclamation mark" and "Comma"
- Visual feedback to indicate speech is being processed

Eye Control

Eye Control makes Windows 10 more accessible by empowering people with disabilities to operate an onscreen mouse, keyboard and text-to-speech experience using only their eyes. The experience requires a compatible eye tracker, like the Tobii 4C, which unlocks access to the Windows operating system to be able to do the tasks one could previously accomplish with a physical mouse and keyboard.

Haptic Feedback

Touch-based feedback usually given as vibration through a controller.

Use Case: Levels of feedback can be altered to provide greater or lesser levels of vibration depending on need or sensitivity.

Keyboard Only

We have enabled the ability to navigate through the Operating System or Office Suite using only the keyboard (no mouse, no touch).

Use Case: Certain users navigate a computer using the keyboard rather than the mouse, as an example, while using a screen reader, working with a Braille keyboard or other Assistive Technology.

Live Tile Sizes

In Windows and Apps, tiles can be changed to multiple sizes for prioritization and easier viewing. Live tiles in Windows 10 display information that is useful at a glance without opening an app.

Use Case: The News tile displays headlines while the Weather tile displays the forecast. You can rearrange, resize and move these tiles to make them work better for you.

Mouse Controls on 10-Key

With mouse controls on 10-key, the number keypad on the keyboard can be used to move the mouse cursor.

Use Case: If someone has limited hand/arm movement staying on the keyboard is extremely useful. It especially helps those with limited shoulder/arm lift.

Office Templates

The most popular templates in Office 365 have accessibility built in. They help you make your content accessible to everyone in your organization.

Want to make your own templates? Several of these templates are not just for you to use, but also guide you through making your own accessible templates. Open the templates to discover tips and how to instructions for making templates for your organization to use.

On-Screen Keyboard

There are several different kinds of keyboards for PCs. The most common type is a physical, external keyboard that you plug into your PC. Windows also has a built-in

Ease of Access tool called the On-Screen Keyboard (OSK) that can be used instead of a physical keyboard to move around a PC's screen or enter text. You don't need a touchscreen to use the OSK. It displays a visual keyboard with all the standard keys, so you can use your mouse, or another pointing device, to select keys or use a physical single key or group of keys to cycle through the keys on the screen.

PowerPoint Designer

PowerPoint Designer improves your slides by automatically generating design ideas that you can choose from. While you're putting content on a slide, Designer works in the background to match that content to professionally designed layouts.

Use Case: People with mobility disabilities and reduced movement of arms/hands can use PowerPoint Designer to help make professional looking slides with fewer clicks of the mouse or keyboard commands.

Remappable Buttons

Mappable buttons allow a game player with a physical disability to have custom buttons for their preferred input to Xbox games. This allows someone who cannot press all the buttons to participate. This is especially powerful when used in conjunction with Co-pilot mode.

Smart Lookup

The Insights pane, powered by Bing, offers more than just definitions. By selecting a word or phrase and launching Smart Lookup, Bing can show you more information, definitions, history and other resources related to that word or phrase.
Select a word or phrase, right-click it and choose Smart Lookup. The insights pane will open with definitions, Wiki articles and top related searches from the web.

Use Case: Smart Lookup allows a person with limited mobility to rapidly access further information from within the Office suite.

Tell Me

The 'tell me what you want to do' feature in the Office ribbon make it easier for you to discover the difficult-to-find capabilities in Office by simply typing into the search box. 'Tell me' will take you directly to the feature/function you want to use without requiring you to search the ribbon or menu.

Use Case: 'Tell me' reduces the need to physically navigate multiple ribbons containing numerous features. This can also be helpful to people who are blind or have low vision, have cognitive disabilities, aging memory loss, or to prompt anyone that can't remember where a specific feature is located.

Text Suggestions in Windows 10

Text suggestions in Windows 10 allows users to get suggested words while they are typing. This is also known as "word prediction." Txt suggestions work with the software keyboard and hardware keyboard and can work with any application in Windows 10.

Themes in Office 365

Themes in Office 365 allow users to pre-set color and text combinations within the Office Suite itself without turning on High Contrast. If you feel like the Office color scheme is too bright or you need more contrast, you can change the Office theme for all your Office programs from your account settings.

Use Case: Changing color schemes can make it easier to see Office 365 applications, potentially reducing eye strain and increasing focus.

Touch Feedback

Touch Feedback shows visual feedback (ripple) when contact is made with a touchscreen and can be darkened for greater visibility.

Use Case: Provides greater focus of where touch contact is made with a screen, useful for more easily tracking contact especially for a person with impaired mobility where fine motor movements may be difficult.

Windows Hello

Windows Hello is a more personal way to sign in to your Windows 10 devices by using facial recognition or your fingerprint. You'll get enterprise-grade security without having to type in a password.

Use Case: Windows Hello provides a secure way to log into a computer for people who easily forget passwords or require assistance entering a password.

Windows Speech Recognition

Windows Speech Recognition lets you control your PC with your voice alone, without needing a keyboard or mouse.

Use Case: Windows speech recognition can be especially helpful for people with disabilities who can't use the keyboard or mouse, but it's available to anyone who'd like to try talking to Windows instead.

ACKNOWLEDGEMENTS

Thanks to Satya Nadella, Frank Shaw and Jason Graefe for encouraging me to report and write The Ability Hacks.

— Greg Shaw

Editorial

Reported and written by Greg Shaw
Cover illustration by Michel Varisco
Reporting assistance by Aileen McGraw
Edited by Aimee Riordan and Roger Ainsley
Special thanks to Julie de Widt-Bakker and Barbara Olagaray Gatto

Eye Gaze – Hackathon

Genevieve Alvarez
Ketura Behrends
Erin Beneteau
Bruce Bracken
Jon Campbell
Eric Chang
Nancy Crowell
Daniel Deschamps
Ashley Feniello

Dave Gaines
Bryan Howell
Tammy King
Jenny Lay-Flurrie
Dan Liebling
Matthew Mack
Rekha Nair
Gershon Parent
Gary Roumanis
Stephen Seed
Mira Shah
Henry Soto
Vidya Srinivasan
Charles Thrasher
Shane Williams
Injy Zarif
Jennifer Zhang

Eye Gaze – With gratitude to our partners

Bill Armstrong
Jeremy Best
Erik Clauson
Sam Dryden
Bob Duffy
Tina Flink
Steve Gleason
Steve Krohn
Kaye Kvam
Father Jim Lee
Tom Lowell

Aaron McQ
Peter Roane
Marc West

Eye Gaze – Post Hackathon

Tambie Angel
Pete Ansell
Eric Badger
Jay Beavers
Bill Buxton
Jon Campbell
Jake Cohen
Ed Cutrell
Maggie Duffield
Alex Fiannaca
Darren Gehring
Austin Hodges
Shaun Kane
Melanie Kneisel
Harish S. Kulkarni
Dwayne Lamb
Rico Malvar
Chuck Needham
Christopher O'Dowd
Ann Paradiso
Narsi Raghunath
Jamie Rifley
Meredith Ringel-Morris

Dmitry Rudchenko
Mira Shah
Noelle Sophy
Irina Spiridonova
Arturo Toledo
Alejandro Toledo
Shane Williams

Learning Tools – Hackathon

Valentin Dobre
Mark Flores
Ryan Galgon
Sebastian Greaves
Greg Hitchcock
Daniel Hubbell
Anna Hughes
Reza Jooyandeh
Kevin Larson
Scott Leong
Tanya Matskewich
Rob McKaughan
Dominik Messinger
Aaron Monson
Pelle Nielsen
Alex Pereira
Jeff Petty
Chris Quirk
Ari Schorr
Stu Shader
Mira Shah

Cynthia Shelly
Mike Tholfsen

Learning Tools – Post Hackathon

Victoria Akulich
Puneet Arora
Yogesh Bhumkar
Sahir Boghani
Tony Chang
Christina Chen
Arnavi Chheda
Clint Covington
Michele Freed
William Fry
Mila Green
Michael Heyns
Dylan Kilgore
Mahesh Kumar
Megan Lawrence
Lan Li
Eliana Liberman
Eran Megiddo
Aldo Navea Pina
Nithin Raj
Priya Rajan
Subha Rao
Malavika
Kyle Ryan
Betty Trevino Sanchez
Yanir Shahak

Guillaume Simonnet
Meng Tan
Ryan Waller
Bo Yan
Yi Zhang

ACCESSIBILITY RESOURCES

Technical

- Developing apps for Accessibility: https://developer.microsoft.com/en-us/windows/accessible-apps
- Microsoft Windows UI Automation Blog: https://blogs.msdn.microsoft.com/winuiautomation/
- Introduction to UIA: Microsoft's Accessibility API: https://www.youtube.com/watch?v=6b0K2883rXA
- Sarah Horton and Whitney Quesenbery. 2014. A Web for Everyone. Designing Accessible User Experiences.
- Matt May and Wendy Chisholm. 2009. Universal Design for Web Applications: Web Applications that Reach Everyone.
- Jonathan Lazar, Daniel Goldstein, and Anne Taylor. 2015. Ensuring Digital Accessibility Through Process and Policy
- Heydon Pickering. Ebook: Inclusive Design Patterns

Disability Culture/Biographies

- Alan Brightman. 2008. DisabilityLand
- Temple Grandin 2009. Thinking in Pictures: My Life with Autism
- Rachel Simons. 2002. Riding the Bus with My Sister
- Lynda Mullaly Hunt. 2015. Fish in a Tree
- Microsoft's Introduction to Disability and Accessibility: https://www.youtube.com/watch?v=GGB_xreE3OU&feature=youtu.be

Disability: Civil Rights

- Joseph P. Shapiro. 1993. No Pity: People with Disabilities Forging a New Civil Rights Movement.
- James I. Charlton. 1998. Nothing About Us Without Us.

Assistive Technology

- Amy G. Dell. 2011. Assistive Technology in the Classroom: Enhancing the School Experiences of Students with Disabilities.

Ted Talks on accessibility

- Design with the Blind in Mind: https://www.ted.com/talks/chris_downey_design_wi th_the_blind_in_mind
- Stella Young: I'm not your inspiration, thank you very much: https://www.ted.com/talks/stella_young_i_m_not_y our_inspiration_thank_you_very_much#t-539327
- Susan Robinson: How I fail at being disabled: https://www.ted.com/talks/susan_robinson_how_i_f ail_at_being_disabled#t-1453
- Sinéad Burke: Why design should include everyone: https://www.ted.com/talks/sinead_burke_why_desig n_should_include_everyone
- Maysoon Zayid: I got 99 problems ... palsy is just one: https://www.ted.com/talks/maysoon_zayid_i_got_99 _problems_palsy_is_just_one
- We all seek the same sense of inclusion: https://www.youtube.com/watch?v=AToBCLl66Ww
- The accessibility equation: valuing an accessible

world for all:
http://tedxauckland.com/videos/accessibility-
equation-valuing-accessible-world_minnie-
baragwanath/

- Judy Heumann's new Ted Talk: Our fight for
  disability rights and why we're not done yet:
  https://www.ted.com/talks/judith_heumann_our_fig
  ht_for_disability_rights_and_why_we_re_not_done_
  yet

Movies on the lives of people with disabilities

- The Silent Child:
  https://www.thesilentchildmovie.com/
- Gleason:
  https://www.imdb.com/title/tt4632316/?ref_=fn_al_tt_1

Made in the USA
San Bernardino, CA
03 July 2018